THE HAND BOOK

THE HAND BOOK

STEPHAN ARIYAN, M.D.

Assistant Professor of Surgery (Plastic)
Plastic and Reconstructive Surgery
Yale University School of Medicine

THE WILLIAMS & WILKINS COMPANY
Baltimore

Copyright ©, 1978
The Williams & Wilkins Company
428 E. Preston Street
Baltimore, Md. 21202, U.S.A.

All rights reserved. This book is protected by copyright. No part of this book may be reproduced in any form or by any means, including photocopying, or utilized by any information storage and retrieval system without written permission from the copyright owner.

Made in the United States of America

Library of Congress Cataloging in Publication Data

Ariyan, Stephan.

  The hand book.

  "A guide to medical students and residents who wish to learn the basic fundamentals of hand surgery."
  Includes bibliographies, references and index.
  1. Hand—Surgery. I. Title. [DNLM: 1. Hand—Surgery. WE830 A718h]
RD559.A73   617'.575   78-13315
ISBN 0-683-00251-1

Composed and printed at the
Waverly Press, Inc.
Mt. Royal and Guilford Aves.
Baltimore, Md. 21202, U.S.A.

Dedication

This book is dedicated to H. Kirk Watson, teacher, Master Surgeon, and life-long friend to each and all of his students of hand surgery. He is Assistant Clinical Professor of Surgery in both the sections of Plastic and Reconstructive Surgery and Orthopaedic Surgery at Yale University School of Medicine, and Chief of the Hand Surgery Service at Hartford Hospital and the University of Connecticut School of Medicine.

His tireless work and dynamic teaching has inspired many a budding hand surgeon. This book was, in fact, conceived as a result of a hand fellowship with him. He shall remain as an ever-present example, guide, critic, and warm friend to all of us.

Stephan Ariyan, M.D.

New Haven, Connecticut
January 1978

Preface

This HAND BOOK is offered as a guide to medical students and residents who wish to learn the basic fundamentals of hand surgery. It is not offered as the "last word" in hand surgery, but to express the feeling that there is no "mystique" about the treatment of the hand. Once the basics are comprehended, hand surgery can be appreciated as an enjoyable and creative field.

Nevertheless, the practice of hand surgery is an experience in continuing education. New techniques and approaches are learned with the passage of time. As such, alternating pairs of pages in this book are left blank for the reader and user to write in these techniques and notes, immediately after the particular topic. In so doing, all the information can be made available to the user when referring back to this book.

Have an enjoyable time.

Acknowledgements

This book is indebted to all those who taught me surgery in general, and hand surgery in particular. There is a tremendous amount of work written on the subject, and many outstanding authors have published the basic knowledge that is brought together in this book. I am particularly grateful to my mentors, Dr. Thomas J. Krizek and Dr. H. Kirk Watson, for the encouragement to write this book.

I am indebted to my associates, Dr. Charles B. Cuono, Dr. Mary H. McGrath, and Dr. Robert Walton for their review, suggestions, and corrections....to Dr. Paul Brown and Dr. Stephen Flagg for their critical evaluation and support....and to my wife, Sandy, for her understanding.

Special appreciation is expressed to my secretary, Helen M. Rogers, for her patience and unfailing energy in typing, and editing, and typing, and editing....The illustrator, Sharon Martin, is to be commended for her work as well as her patience with the repeated changes that were necessary. Much of the artwork has been based on the works of those hand surgeons who have contributed so much to the furthering of the knowledge in this area. Credit is given to them in general in the text, and in particular on page 292.

The efforts and support of all these people have made the writing of this book a pleasurable experience.

Contents

	Preface	vii
	Acknowledgments	ix
I.	Anatomy	1
II.	Examination	61
III.	Incisions	77
IV.	Splinting	85
V.	Common Injuries and Deformities	93
VI.	Flexor Tendon Injuries	173
VII.	Extensor Tendon Injuries	201
VIII.	Nerve Repairs	209
IX.	Tendon Transfers	221
X.	Rheumatoid Arthritis	245
XI.	Degenerative Arthritis	257
XII.	Congenital Deformities	277
	Artwork Acknowledgments	292
	Bibliography	294
	Index	299

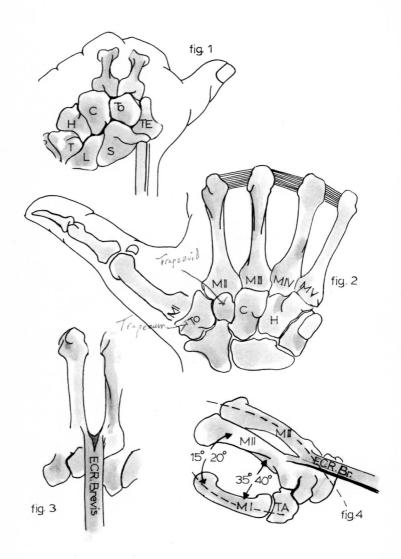

I

Anatomy

1. BONE

   The structure and architecture of the hand is based on the bones.[1] The CARPAL bones form a semicircular arc. The CAPITATE is the "Keystone" (Fig. 1), and is flanked on the radial side by the TRAPEZOID and TRAPEZIUM and on the ulnar side by the HAMATE.

   The two central metacarpals (M_{II} and M_{III}) are bound firmly to the Trapezoid and Capitate (Fig. 2) and constitute the FIXED UNIT of the hand.

   There are 3 wrist extensors: ECRL, ECRB, ECU. The centrally located ECRB is the PRIME wrist extensor (Fig. 3) because of its insertion on the base of the FIXED UNIT allowing the flexors to act through their range on a MECHANICALLY STABLE wrist.

   Therefore, DO NOT USE ECRB FOR TRANSFERS, and DO NOT FUSE THE WRIST EXCEPT FOR CHRONIC PAIN, INSTABILITY, or DEFORMITY.

   Thumb

   There is a unique functional relationship of the CMC joint to the other fingers. The articulation of the M_I to the TRAPEZIUM is that of a "SADDLE JOINT". As such, the M_I can have a range of motion through the CMC joint in 3 planes:

   A_1 = Flexion-Adduction/Extension-Abduction
   A_2 = Adduction/Abduction (Palmar plane)
   A_3 = Circumduction

   However, intrinsic muscle support is necessary for the stability of this motion. If stability is lost, then the CMC joint must be fused to allow opposition of the thumb.

NOTES

NOTES

4 THE HAND BOOK

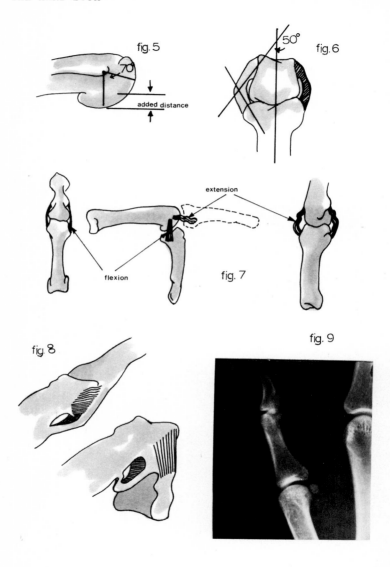

fig. 5 added distance

fig. 6 50°

fig. 7 extension flexion

fig. 8

fig. 9

Therefore, CMC FUSION SHOULD BE DONE IN OPPOSITION (Fig. 4):

    15-20° radial abduction
    35-40° palmar abduction

2. JOINTS

Metacarpophalangeal

    The head of the metacarpal bone has an OVOID (eccentric) surface in a sagittal plane resulting in a CAM effect (Fig. 5) and a trapezoid shape in the cross-sectional plane (with volar flare) (Fig. 6).[2]

    The COLLATERAL ligaments are somewhat TRIANGULAR in shape and arise DORSAL to the axis of rotation of the metacarpal head (Fig. 7). As a result of the CAM effect and VOLAR FLARE, the collateral ligaments are stretched and tightened on flexion, and relaxed and loosened on extension (Fig. 7). Therefore, the MP joint is stable in flexion and cannot abduct, while it is loose in extension allowing lateral motion and abduction-adduction.

    The MP joints are STABLE IN FLEXION AND MOBILE in EXTENSION.

Interphalangeal

    The IP joint, on the other hand, has the collateral ligaments originating CENTRAL in the axis of rotation. It is the radius of a CONCENTRIC arc, with only the central fibers in stretch at all times. As such, in full EXTENSION, the upper fibers are loose while the lower fibers are stretched; in full FLEXION, the upper fibers become taut while the lower fibers remain tight as they move over the condylar flare (Fig. 8).

    The IP joints are STABLE IN ALL POSITIONS. This is important for stable pinch.

Thumb

    The IP joint of the thumb is basically identical to the IP joint of the fingers. The only exception is that it has ONE SESAMOID (Fig. 9).

NOTES

NOTES

8   THE HAND BOOK

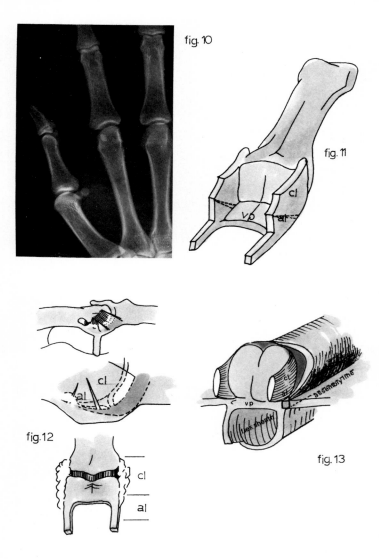

fig. 10

fig. 11

fig. 12

fig. 13

The MP joint of the thumb is a transition between a digital IP and MP. In some it has a CONCENTRIC head, and in others it has a SQUARE head. It has 2 SESAMOIDS for insertions of intrinsic muscles (Fig. 10).

RADIAL is the insertion of the lateral head of FPB.

ULNAR is the insertion of ADDUCTOR POLLICIS.

3. LIGAMENTS

For purposes of discussion, the IP joint and its support structures can be thought of geometrically as a BOX (Fig. 11).[2] The front and back walls are the ARTICULAR SURFACES of the joint.

The floor is made of the VOLAR PLATE, a fibrocartilaginous structure that is thick at its insertion distally at the base of the phalanx, and thinned proximally so that it may fold on itself on flexion of the joint (Fig. 8c).

The thicker COLLATERAL ligaments are fibers that originate at the head of the condyle and insert distally at the base of the articulating phalanx (Fig. 12). The thinner ACCESSORY collateral ligaments are made of fibers that originate at the condylar head and insert on the volar plate. In reality, these fibers comprise a continuous sheet with no true separation and only artificially designated as "collateral" and "accessory" by the location of their insertions.

At the more proximal portion, the accessory ligament and the volar plate blend with the FLEXOR TENDON SHEATH as they all insert on the periosteum of the phalanx at the "ASSEMBLY LINE" (Fig. 13). This blending of the three structures form the CHECK-REIN ligaments, and can be thought of as the handles of a wheelbarrow (Fig. 11). These check-rein ligaments restrain the joint from going into hyperextension, yet are flexible enough to fold and permit flexion (Fig. 8c).

NOTES

NOTES

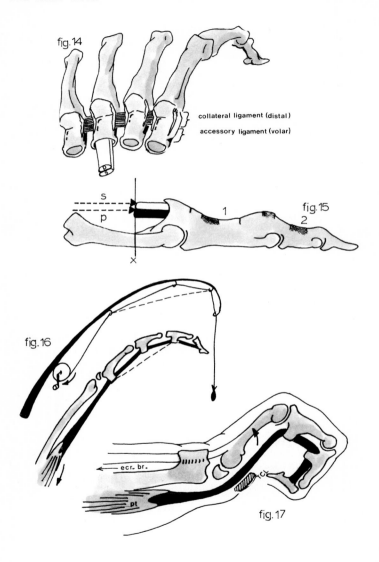

At the MP joint, the TRANSVERSE METACARPAL LIGAMENTS bind together the blending of the volar plates and flexor sheaths and allow the insertions of the accessory collateral ligaments (Fig. 14). As described earlier, the collateral ligaments are stretched on flexion and relaxed on extension, allowing stability in flexion and lateral mobility in extension.

AT THE MP JOINT, IF THE LIGAMENTS ARE TORN OR CUT, LATERAL STABILITY MAY BE MAINTAINED BY THE INTEROSSEI, SUPPORTED BY THE FLEXOR SHEATHS.

AT THE IP JOINTS, THE COLLATERAL LIGAMENTS ARE THE ONLY STABILITY AND AT LEAST 2 ELEMENTS MUST BE LOST TO LOSE LATERAL STABILITY.

Fibro-osseous Tunnel

The FLEXOR TENDON SHEATH extends from the head of the metacarpals to the insertion of the flexor tendon at the distal phalanx (Fig. 15). Along this sheath, there are two areas of prominent thickenings or ANNULAR BANDS that need to be preserved to act as pulleys. These two important thickenings are located at the proximal portion of the proximal phalanx, and the middle portion of the middle phalanx. A third, though less important pulley, is located at the proximal opening of the sheath. These pulleys serve the same function as the eyelet loops of a fishing rod (Fig. 16). They prevent bowstringing across the arc of the rod (phalanges), and in their absence, the excursions and force of the line (tendon) is wasted.

PRESERVE OR RECONSTRUCT THE PULLEYS AT THE PROXIMAL AND MIDDLE PHALANGES.

Transverse Carpal Ligament

In addition to the pulley mechanisms of the flexor tendon sheath, the TRANSVERSE CARPAL LIGAMENT (TCL) spans the volar aspect of the proximal palm to form the roof of the CARPAL TUNNEL (Fig. 17). The TCL traverses from the SCAPHOID tubercle and crest of the TRAPEZIUM on the radial side to the PISIFORM and HAMATE on the ulnar side (Fig. 18). Also on the ulnar side of the wrist, the CANAL OF GUYON, or ulnar tunnel, is formed by the TCL along the floor, the VOLAR CARPAL LIGAMENT along the roof, and the PISIFORM along the ulnar wall (Fig. 18).

NOTES

NOTES

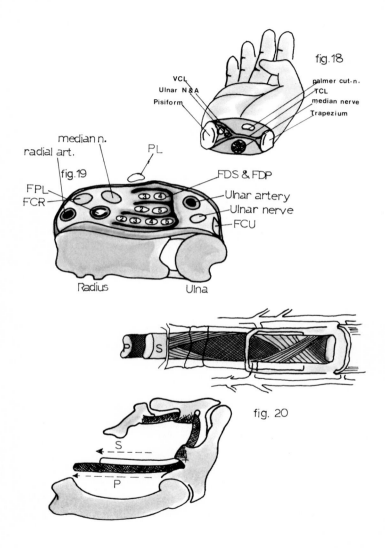

4. TENDONS

Wrist Tendons

The FCR runs along the radial margin of the TCL, through its own sheath, to insert into the bases of the M_{II} M_{III}. The FCU inserts into the PISIFORM and base of M_V. PL inserts into the aponeurosis of the PALMAR FASCIA.

The extensors counterbalance these by insertions basically at the same points. The ECRL inserts at the base of M_{II}, the ECRB at the base of M_{III} and the ECU at the base of M_V.

AS MENTIONED BEFORE, ECRB INSERTS AT THE BASE OF THE FIXED UNIT AND IS THE PRIME EXTENSOR.

Digital Flexor Mechanism

The FPL lies along the radial side of the carpal tunnel and inserts at the base of the distal phalanx of the thumb.

The digital flexor tendons at the wrist, as they enter the carpal tunnel, are arranged with the deep flexors lying side by side along the floor. The superficial flexors are arranged volar to these in two layers: the tendons to the LONG and RING fingers (3rd and 4th) are volar to the tendons to the INDEX and LITTLE (2nd and 5th). Remember 34 is a higher number than 25 (Fig. 19). In the palm, each of the FDS lies superficial to the FDP of the corresponding finger. As this pair of tendons enter the fibro-osseous tunnel of each finger, the FDS divides into the two slips, allowing the FDP to pass through and lie more superficial, distal to the base of the proximal phalanx (Fig. 20). As the FDS moves distally it re-deccusates and splays out to insert at the middle portion of the middle phalanx. The FDP continues distally to insert along the volar aspect of the midportion of the distal phalanx.

NOTES

NOTES

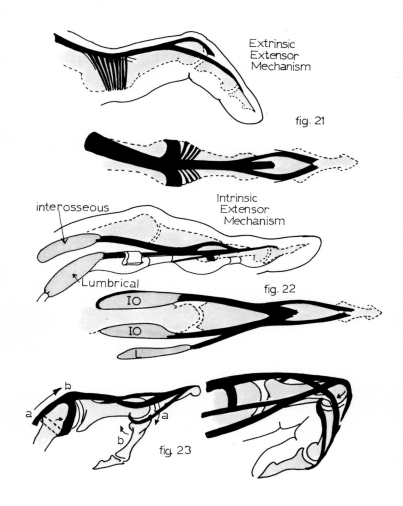

Digital Extensor Mechanism

The digital extensor mechanism is a CONJOINED tendinous structure made up of the EXTRINSIC long extensor tendon and the tendinous insertions of the INTRINSIC muscles. The EXTRINSIC long extensor portion has 4 components (Fig. 21).

1. The SAGITTAL BANDS (shroud ligaments) encircle the head of the metacarpal.

2. The PROXIMAL SLIP inserts into the base of the proximal phalanx.

3. The CENTRAL SLIP inserts into the base of the middle phalanx.

4. The LATERAL SLIPS insert at the base of the distal phalanx.

The INTRINSIC MUSCULAR portion has tendinous extensions from the volar and dorsal interossei to the radial and ulnar lateral slips of the extrinsic portion, and extensions of LUMBRICAL muscles to the RADIAL lateral slips of the extrinsic portion (Fig. 22). These split to send one slip to insert with the central slip into the middle phalanx, and one slip to join the lateral slip of the long extensor to insert into the base of the distal phalanx.

THE EXTENSOR MECHANISM IS A COMBINATION OF THE EXTRINSIC AND INTRINSIC PORTIONS.

On flexion of the MP joint, the sagittal bands SLIDE FORWARD to apply the extensor forces along the central slip and extend the middle phalanx (Fig. 23). On flexion of the PIP joint, the lateral slips slide volar to the axis of rotation to loosen and allow flexion of the DIP joint. Otherwise, it would be taut and extend the DIP by a tenodesis effect.

The EXTENSOR HOOD mechanism is a triangular sleeve extending from the head of the metacarpal to the DIP joint and is made up of (Fig. 24):

NOTES

NOTES

24 THE HAND BOOK

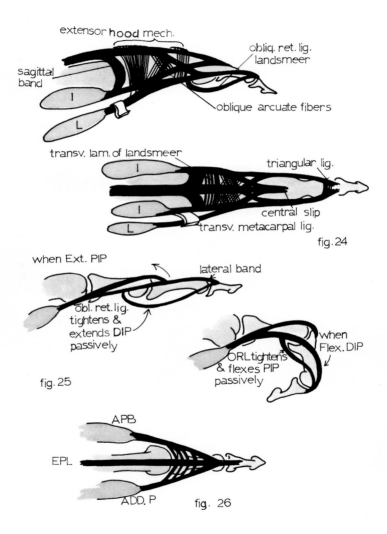

fig. 24

fig. 25

fig. 26

1. OBLIQUE ARCUATE FIBERS criss-crossing between the central slip and the lateral bands.
2. TRANSVERSE LAMINA OF LANDSMEER traversing over the dorsum of the joint between the assembly lines.
3. TRIANGULAR LIGAMENT extending between the insertions of the lateral bands on the distal phalanx.
4. OBLIQUE RETINACULAR LIGAMENT OF LANDSMEER which stretches from the extensor insertion in the distal phalanx over the axis of the PIP joint and on to the assembly line (Fig. 13). It acts to COORDINATE the uniform flexion and extension of the joints.[3] As the FDP flexes the DIP joint, the ligament tightens and flexes the PIP joint passively through a tenodesis effect (Fig. 25). Similarly, as the extensors extend the PIP joint, the ligament is again put on stretch, which helps extend the DIP joint. However, this dynamic function of the ligament may be in dispute, for careful studies have found it to be present in only 50% of the fingers examined.[4]

Thumb Extensor Mechanism

The extensor mechanism of the thumb is a modified form of the digital mechanism, since there is no middle phalanx. The EXTRINSIC portion is made up of the EPB which runs along the middle of the dorsum of the MP joint to insert on the proximal phalanx, and the EPL which runs over the ulnar side of the dorsum of the MP joint to insert into the base of the distal phalanx (Fig. 26).

The INTRINSIC portion is made up of the APB which sends fibers to insert on the radial side of the proximal phalanx then over the dorsum of the joint to the two extensor tendons and meeting fibers from the ADD.P. from the ulnar side of the phalanx.

ALTHOUGH THE THUMB HAS NO LATERAL BANDS, THIS DORSAL FIBROUS EXPANSION ACTS AS AN EXTENSOR HOOD, WITH DISTINCT EDGES MADE UP OF THE EXPANSION OF THE APB AND ADD.P. WHICH ARE THE "LATERAL BAND EXTENSORS".

THE IP JOINT OF THE THUMB IS EXTENDED BY THE ACTIONS OF ALL THREE NERVES: MEDIAN (APB), RADIAL (EPL), ULNAR (ADD.P.).

NOTES

NOTES

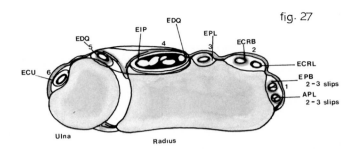

fig. 27

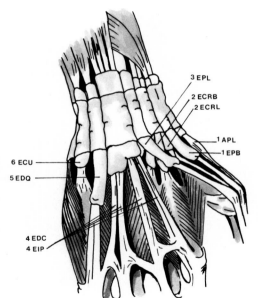

fig. 28

Extensor Compartments At The Wrist

The extensor tendons pass through 6 COMPARTMENTS under the EXTENSOR RETINACULUM. From radial to ulnar they can be remembered by the numbers "221211" (Fig. 27).

2   ABDUCTOR POLLICIS LONGUS
    EXTENSOR POLLICIS BREVIS

2   EXTENSOR CARPI RADIALIS LONGUS (inserts on M_{II})
    EXTENSOR CARPI RADIALIS BREVIS (inserts on M_{III})

1   EXTENSOR POLLICIS LONGUS

2   EXTENSOR DIGITORUM COMMUNIS
    EXTENSOR INDICIS PROPRIUS

1   EXTENSOR DIGITI QUINTI PROPRIUS

1   EXTENSOR CARPI ULNARIS (inserts on M_V)

It should be noted that the "SNUFFBOX" is made up of the space between the EPB and EPL (Fig. 28). Also, the ECRL and ECRB pass through this snuffbox as they insert into the metacarpals.

THE TWO INDEPENDENT EXTENSOR TENDONS, EIP AND EDQ, BOTH LIE ULNAR TO THE COMMON EXTENSOR TENDON NEAR THE METACARPAL HEADS.

The EDQ has two tendinous slips. Finally, an important anatomic consideration in the first compartment to be remembered in DeQuervain's Disease is that the APL AND THE EPB MAY HAVE MORE THAN ONE TENDON OR TUNNEL IN THE FIRST COMPARTMENT (Fig. 27).

The APL usually has 2-3 slips, the first 1 or 2 going to the APB and an additional one going to the M_I. The EPB may also have 2-3 slips. When releasing this compartment, all these tunnel sheaths need to be opened.

NOTES

NOTES

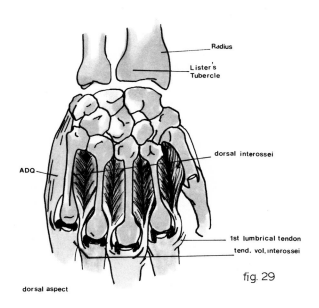

fig. 29

dorsal aspect

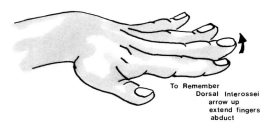

To Remember
Dorsal Interossei
arrow up
extend fingers
abduct

5. MUSCLES
   A. EXTRINSIC
      Dorsal (Fig. 28):
         Brachioradialis (BR)
         Abd. Poll. Long. (APL)
         Ext. Poll. Brev. (EPB)
         Ext. Carp. Rad. Long. (ECRL)
         Ext. Carp. Rad. Brev. (ECRB)
         Ext. Poll. Long. (EPL)
         Ext. Dig. Comm. (EDC)
         Ext. Ind. Prop. (EIP)
         Ext. Dig. Quinti (EDQ)
         Ext. Carp. Uln. (ECU)

      Volar (Fig. 19):
         Pronator Teres (PT)
         Palmaris Long. (PL)
         Flex. Carp. Rad. (FCR)
         Flex. Poll. Long. (FPL)
         Flex. Dig. Subl. (FDS)
         Flex. Dig. Prof. (FDP)
         Flex. Carp. Uln. (FCU)

NOTES

NOTES

36 THE HAND BOOK

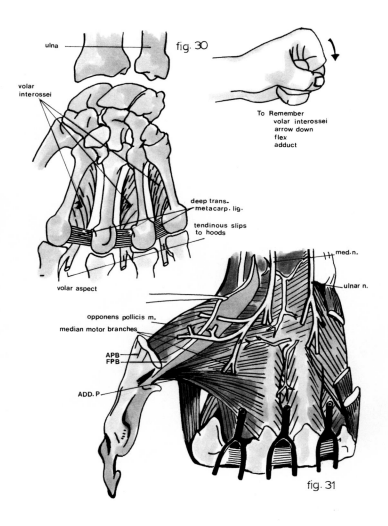

ulna

fig. 30

volar interossei

To Remember
volar interossei
arrow down
flex
adduct

deep trans-
metacarp. lig.

tendinous slips
to hoods

volar aspect

opponens pollicis m.
median motor branches
APB
FPB
ADD. P

med. n.
ulnar n.

fig. 31

B. INTRINSIC

Dorsal Interossei (DI) - 4 in number, arising from the adjacent surfaces of the shafts of the M_{I-V} and insert on the lateral slips of the extensor hood of M_{II-IV} (Fig. 29). They (ABDUCT) move M_{II} radialward, M_{IV} ulnarward and M_{III} both radial and ulnarward. They lie dorsal to the TMCL.

Volar Interossei (VI) - 3 in number, arising from $M_{II,IV,V}$ and inserting on the lateral slip of their respective extensor hood mechanisms (Fig. 30). They (ADDUCT) move $M_{IV,V}$ radialward, and M_{II} ulnarward. They lie dorsal to the TMCL.

Lumbrical - 4 in number (Fig. 31). The first and second lumbricals arise from the radial side of their respective FDP tendons. The third arises from the adjacent sides of the 2nd and 3rd FDP, and the fourth arises from the adjacent sides of the 3rd and 4th FDP.* They all lie VOLAR to the TMCL and insert on the RADIAL aspect of the lateral slips of their respective extensor hood mechanisms.

Thenar Muscles (Fig. 31):

    ABD. Poll. Brev. (APB)

    Opponens Pollicis (OP)

    Flex. Poll. Brev. (FPB)

    Adductor Pollicis (ADD.P.)

Hypothenar Muscles (Fig. 31):

    ABD. Dig. Quinti (ADQ)

    Flex. Dig. Quinti (FDQ)

    Opponens Dig. Quinti (ODQ)

---

*See "Examinations", p. 61.

NOTES

NOTES

40   THE HAND BOOK

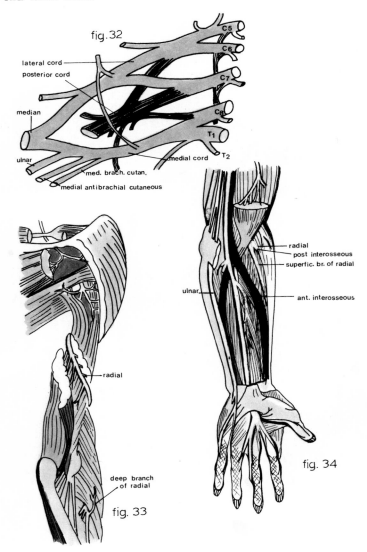

fig. 32
lateral cord
posterior cord
C5
C6
C7
C8
T1
median
T2
ulnar
medial cord
med. brach. cutan.
medial antibrachial cutaneous

radial
deep branch of radial
fig. 33

radial
post interosseous
superfic. br. of radial
ulnar
ant. interosseous
fig. 34

6. NERVES

   A. Radial N. - arises from posterior cord of the brachial plexus (Fig. 32), it spirals down the humerus (Fig. 33) to lie under the BR on the ulnar side of the arm and is subject to injury from fracture of the humerus.

   Above the elbow it gives branches to the BRACHIORADIALIS (Fig. 38), at the elbow it divides into the SUPERFICIAL BRANCH OF THE RADIAL N. and the POSTERIOR INTEROSSEUS N. with a branch to the SUPINATOR.

   (1) SUPERFICIAL BRANCH of the Radial N. supplies the ECRL, ECRB then continues along the radial border of the wrist to supply the skin over the radial portion of the dorsal hand (Fig. 37).

   (2) POST. INTEROSSEUS N. supplies the APL, EPB, EPL, EDC, EIP, EDQ, ECU (Fig. 34).

   B. Median N. - arises from lateral and medial cords of the brachial plexus and comes down the arm with the brachial artery, to be divided high in the forearm at the heads of the PRONATOR TERES, giving off the...

   (1) ANTERIOR INTEROSSEUS N. (Fig. 34) which supplies the FPL and FDP to the index and long fingers, and PRONATOR QUADRATUS.

   (2) The main branch of the MEDIAN N. descends vertically behind the FDS to lie just beneath and radial to the PL at the wrist. As it passes through the carpal tunnel, it divides into 5 branches: a MEDIAN MOTOR Branch, and 4 DIGITAL Branches (Fig. 35).

   The Median motor nerve may pass around the distal end of the TCL, or may pierce the TCL to supply the APB, OP, and superficial head of FPB (Figs. 31 & 35).

NOTES

NOTES

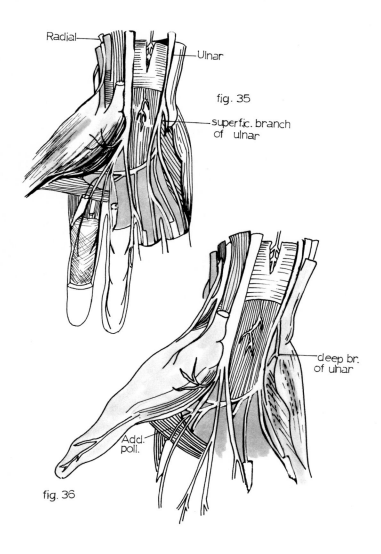

fig. 35

fig. 36

The 4 DIGITAL branches supply sensation to the thumb, radial side of the index, adjacent sides of the index and long fingers, and the radial side of the ring finger. The 2nd and 3rd digital branches also give off motor branches to the 1st and 2nd LUMBRICAL muscles (Fig. 35).

C. Ulnar N. - comes down the arm and enters the forearm below the elbow by descending between FDS and FCU, where it gives a branch to the FCU and the FDP to the ring and little fingers (Fig. 34). It descends down the forearm along the ulnar side of the ulnar artery to pass through the CANAL of GUYON and exits by dividing into SUPERFICIAL and DEEP branches.

   (1) The SUPERFICIAL BRANCH divides into two digital branches supplying sensation to the little finger and ulnar half of the ring finger (Fig. 35).

   (2) The DEEP BRANCH supplies motor fibers to the ADQ, FDQ, ODQ, then follows the deep ulnar arterial arch to supply fibers to the 3rd and 4th LUMBRICALS, DORSAL and VOLAR INTEROSSEI, DEEP HEAD of the FPB, and ends at the ADDUCTOR POLLICIS (Fig. 36).

NOTES

NOTES

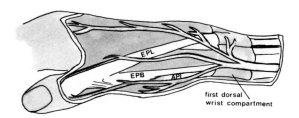

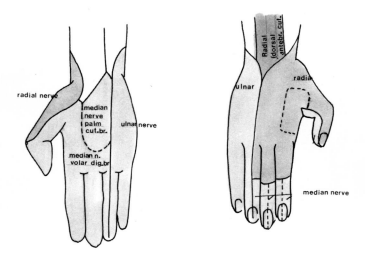

AREAS OF SKIN INNERVATION

fig. 37

OUTLINE OF NERVE SUPPLIES TO MUSCLES

	RADIAL		MEDIAN	ULNAR
	BR		PT	FCU
	SUPINATOR		FCR	FDP (3,4)
	ECRL		PL	ADQ
	ECRB		FDS	FDQ
	APL		FPL	ODQ
Post. Interosseus N.	EPB	Ant. Inteross. N.	FDP (1,2)	LUMB (3,4)
	EPL		PRON. QUAD.	DI
	EDC		APB	VI
	EIP		OP	FPB ($\frac{1}{2}$)
	EDQ		FPB ($\frac{1}{2}$)	ADD.P.
	ECU		LUMB (1,2)	

NOTES

NOTES

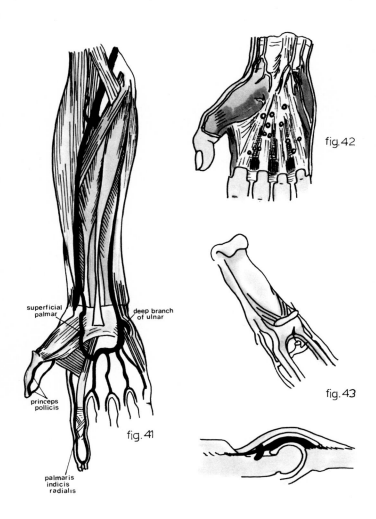

fig. 41
superficial palmar
deep branch of ulnar
princeps pollicis
palmaris indicis radialis

fig. 42

fig. 43

B. ULNAR artery descends down the ulnar side of the forearm (Fig. 39) traverses the CANAL of GUYON lying radial to the ulnar nerve and divides into the larger SUPERFICIAL and smaller DEEP branches. The DEEP PALMAR branch supplies the hypothenar muscles and forms the DEEP VOLAR ARCH with the larger terminal branch of the radial artery. The larger SUPERFICIAL branch forms the SUPERFICIAL VOLAR ARCH, communicating with the smaller superficial branch from the radial artery, and providing the DIGITAL arteries to the ulnar side of the index, and both sides of the middle, ring, and little fingers (Fig. 41).

From this distribution, note that the thumb has good blood supply to the dorsal (1st DORSAL METACARPAL) as well as the volar skin (PRINCEPS POLLICIS). THIS PERMITS THE ELEVATION AND ADVANCEMENT OF VOLAR (MOBERG) SKIN FLAPS IN AMPUTATED TIPS.

However, the remaining fingers have their blood supply mostly from the DIGITAL vessels (off the SUPERFICIAL PALMAR ARCH) which send blood to the dorsal skin, and usually lesser amounts from the DORSAL METACARPAL vessels. THIS MAKES A LARGE VOLAR (MOBERG) ADVANCEMENT FLAP MORE RISKY IN THE OTHER FINGERS.

THE BLOOD SUPPLY TO THE SKIN OF THE PALM IS IN A VERTICAL DIRECTION FROM THE PALMAR ARTERIAL ARCH (Fig. 42).

The arterial blood supply to the thumb and radial index digital come from the DEEP PALMAR ARCH (radial artery) while the ulnar index digital and remaining fingers get their arterial supply from the SUPERFICIAL PALMAR ARCH (ulnar artery).

The two digital vessels on each side of a finger communicate with each other by means of a TRANSVERSE BRANCH that is deep to the CHECK-REIN ligaments (Fig. 43). Therefore, when RELEASING THE VOLAR PLATE OR CHECK-REIN LIGAMENTS TAKE CARE NOT TO DAMAGE THIS COMMUNICATIONS VESSEL.

NOTES

NOTES

60 THE HAND BOOK

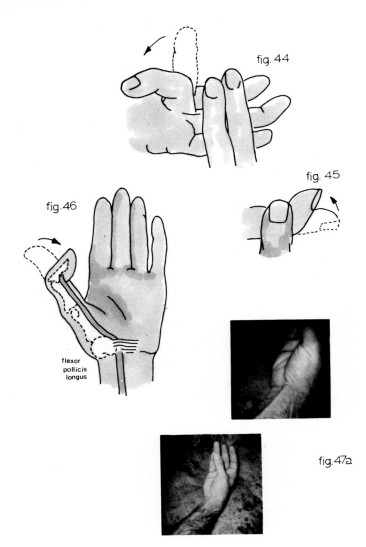

fig. 44

fig. 45

fig. 46

flexor
pollicis
longus

fig. 47a

II

Examination

    Examination of the hand should be done methodically. If the hand has sustained trauma, the skeletal structures should be examined physically and radiographically for fractures. The integrity of the tendons or the continuity of nerves to muscles can be demonstrated by the selected motions of the fingers, wrist, and forearm.

    Joints and Ligaments - must be tested by stress. The IP joints need to be stressed laterally. The MP joints have lateral stability only in flexion and must be tested in this position (Fig. 7).

Tendons

FDS　　　　　　- flexes PIP joint. Test finger by stabili-
(Fig. 44)　　　　zing remaining fingers in extension against
　　　　　　　　a table top.

FDP, FPL　　　- flexes DIP joint. Test finger by stabili-
(Fig. 45)　　　　zing middle phalanx in extension against a
(Fig. 46)　　　　table top.

DI, VI　　　　　- abduct-adduct fingers. Test by asking
　　　　　　　　patient to adduct and abduct with fingers
　　　　　　　　in extension.

PL
(Fig. 47A)　　　- to check for the presence of a palmaris
　　　　　　　　longus for a tendon transfer, ask the
　　　　　　　　patient to flex the wrist while opposing
　　　　　　　　thumb to little finger. This tendon may be
　　　　　　　　absent in 20% of the population.

Nerves

    Sensory portion of the nerve may be examined for by 2-point discrimination with a paperclip or a calipre. Normal sensory 2-point discrimination is up to 5mm at the volar aspect of the fingertips and much wider elsewhere.

NOTES

NOTES

64   THE HAND BOOK

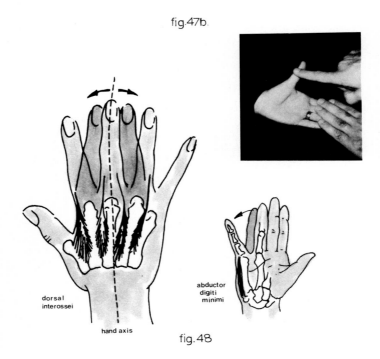

fig. 47b.

dorsal
interossei

hand axis

abductor
digiti
minimi

fig. 48

fig. 49a

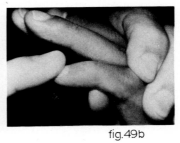

fig. 49b

The median and ulnar nerve distributions have been described earlier (Fig. 37). The radial nerve is best examined for at the dorsal skin of the thumb-index web space.

Motor portion may be examined as follows:

Radial N. - wrist extension

Median N. - thumb opposition to vertical plane of mid palm with back of extended hand resting on table top (Fig. 47B).

Ulnar N. - abduction-adduction of fingers. The index finger is abducted by the 1st DI and the thumb is adducted by the ADD.P., both of which are supplied by the TERMINAL portion of the MOTOR BRANCH of the ulnar nerve, which would demonstrate integrity of the nerve more proximally (Fig. 48).

Compromise of these nerves may be determined by examining strength of these muscles against resistance by the examiner.

Tests

Allen - tests circulation through the radial and ulnar arteries at the wrist. With the patient making a tight fist, occlude both vessels with examining fingers. Then release one or the other vessel and check for capillary refill of blanched palmar and digital skin.

Boutonniere - check passive flexion of DIP joint while PIP is in flexion. Then, with PIP held in resisted extension with one examining hand, check resistance of passive flexion of DIP joint (Fig. 49A). This test reflects tightness in the lateral bands due to volar displacement and shortening.

Bunnel or Fowler - intrinsic tightness test. Check passive flexion of DIP joint while MP and PIP are partially flexed. Then, with MP held in resisted extension with one examining hand, check resistance of passive flexion of DIP joint (Fig. 49B). Test reflects tightness of lateral bands 2° to intrinsic tightness.

Finkelstein - with thumb flexed into palm of fist, actively or passively deviate wrist ulnarward with wrist in neutral. Stenosing tenosynovitis of first dorsal compartment of wrist (APL and EPB) will cause pain.

NOTES

NOTES

68   THE HAND BOOK

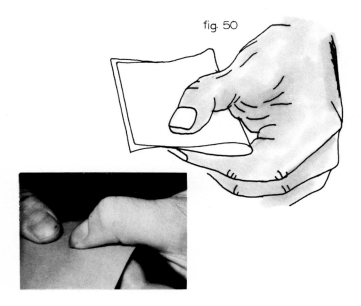

fig. 50

fig. 51

Froment Sign - describes loss of thumb adduction due to ulnar nerve loss and loss of pinch due to 1st DI and ADD.P. Substitutes with EPL because it goes around LISTER'S TUBERCLE and adducts. Cannot abduct index either, so they pinch with thumb to index MP. They use FPL to flex and STABILIZE thumb, and EPL to ADDUCT by straightening line (Fig. 50). [1,7]

Tinel - Tapping along a nerve will cause distal paresthesias over areas of nerve regeneration. Can test degree of compromise or regeneration. Very subjective!

Phalen - with the elbows resting on table top, passively flex both wrists fully (Fig. 51). Paresthesias or hyperthesias along sensory distribution of median or ulnar nerves within 60 seconds reflects compromise in these nerves.[8]

Nerve Blocks

Sometimes examination of traumatic wounds are facilitated following a nerve block.

Wrist

Median N. - inject under skin proximal to wrist crease and radial to PL (located by thumb-index opposition and wrist flexion) (Fig. 52).

Ulnar N. - inject under skin proximal to wrist crease and radial to FCU (located by abducting little finger and flexing wrist). May need to infiltrate skin on ulnar side of dorsal carpal area to get dorsal cutaneous branches (Fig. 53).

Radial N. - inject under skin in "snuffbox" between EPB and EPL (located by extending thumb) (Fig. 54).

NOTES

NOTES

72 THE HAND BOOK

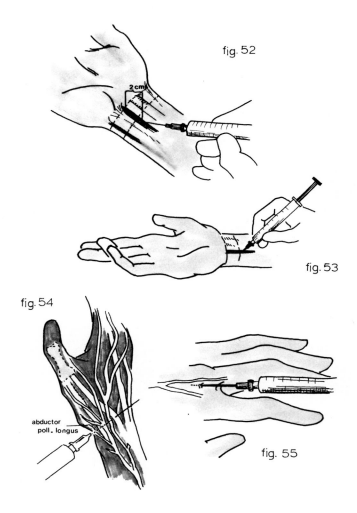

fig. 52

fig. 53

fig. 54

abductor
poll. longus

fig. 55

Digital

(a) finger blocks may be achieved by injecting the web space with a 3/4" - 25 gauge needle in a horizontal plane the full length of the needle (Fig. 55).

(b) may also be achieved by injecting on both sides of metacarpal heads at level of distal palmar crease at palm.

(c) last choice is "ring block" by injecting the digital nerves on both sides of proximal phalanx, and dorsal skin. May cause damage to neurovascular bundle with needle, or compression occlusion of blood supply by fluid volume.

NEVER USE EPINEPHRINE IN DIGITAL OR WRIST BLOCKS. UNNECESSARY TO USE MORE THAN 2-5 CC OF 1-2% XYLOCAINE.

NEVER BLOCK NERVES UNTIL PROPER NEUROLOGIC EXAMINATION IS PERFORMED AND RECORDED.

Cleansing

Wounds seen in emergency rooms often are dirty with greasy and tarry products. These may very readily be cleansed by applying a topical antibiotic ointment (e.g., Neosporin® etc.) which has a petrolatum base to dissolve the grease and tar and remove it. Several applications and wipings may be necessary to cleanse the wound. Sterile toothbrushes kept in the emergency room are very useful for cleansing of wounds after they have been anesthesized. A jet-pulsatile lavage is indispensable in removing foreign bodies.[9] If one is not available, a 50 cc syringe filled with saline may be used to produce a strong enough pressure of fluid for similar debridement.[10]

NOTES

NOTES

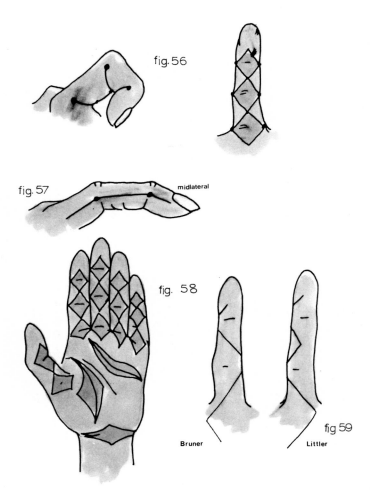

fig. 56

fig. 57   midlateral

fig. 58

fig 59

Bruner   Littler

III

Incisions

Skin incisions must be planned in a manner to avoid contractures. This is achieved by not crossing flexion creases. The areas of palmar and volar digital skin that come in contact with each other on flexion must not be crossed (Fig. 56). Skin incisions along the flexion creases are safe.

Incisions

Mid axial — marking the dorsal extent of flexion crease of each IP joint and marking these. This line is dorsal to the NV bundle. These lines do not change dimension on flexion or extension (Fig. 57).

Bruner — zig-zag incision of volar digital skin crossing between lateral points of flexion creases (Fig. 58).[11]

Littler — modification of Bruner incision by incompletely crossing the volar skin (Fig. 59).

Z-plasty — double-opposing transposition flaps (Fig. 60) serve four purposes:

    (a) increase length of skin

    (b) change direction of scar

    (c) form web space

    (d) change position of topography (corner of lip or eye)

— remember, a z-plasty gives greater length at the expense of shorter width. A four-flap z-plasty may be very useful in lengthening of a thumb-index web contracture (Fig. 61).

NOTES

NOTES

80 THE HAND BOOK

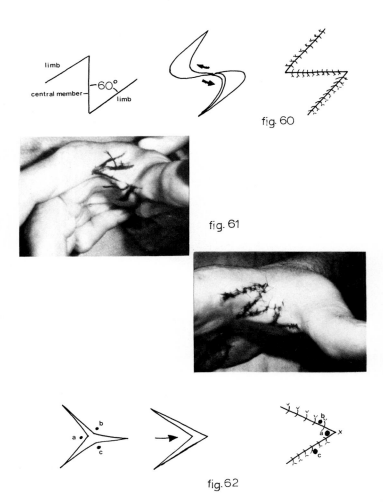

fig. 60

fig. 61

fig. 62

Y-V-plasty   -   along the horizontal plane of palm or
                     finger increases longitudinal length
                     without complete undermining or elevation
                     of skin from underlying structures and
                     blood supply (Fig. 62).[12]

   The use of both z-plasty and y-v-plasty are very helpful in the treatment of Dupuytren's contracture (see p. 149).

NOTES

NOTES

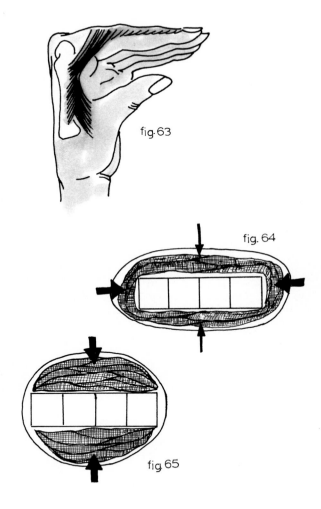

fig. 63

fig. 64

fig. 65

IV

Splinting

Dressings perform several functions:
1. immobilization
2. compression
3. absorbtion
4. medication - drugs, moisture, etc.
5. antisepsis
6. aesthetic
7. comfort

As such, the first four listed above, are the most common indications for splinting a hand.

The proper position to immobilize a hand is with WRIST in dorsiflexion, MP at 90°, IP in extension, and THUMB in abduction. This is called the "POSITION OF SAFETY", because it is the safest position from which stiffness can be treated to return to normal function (Fig. 63).

Consider the tendons as wet spaghetti: you can pull it along a plane, but you cannot push it!

In the POSITION OF SAFETY, the extensor tendons and flexor tendons are mid-way in position. Adhesions of the extensor can be broken by flexion, and of the flexors by extension. In this POSITION, the intrinsics are the most relaxed, and can be stretched by extending the MP and flexing the IP.

An occlusive dressing can be applied by placing fluffs in the palm and dorsum of the hand and not CIRCUMFERENTIALLY. Padding circumferentially around a rectangle (four fingers) gives more compressions along the sides of the index and little finger rays, and less in the palm and dorsum (Fig. 64). With padding along the palm and dorsum only, more forces are applied in this direction (Fig. 65).

NOTES

NOTES

88 THE HAND BOOK

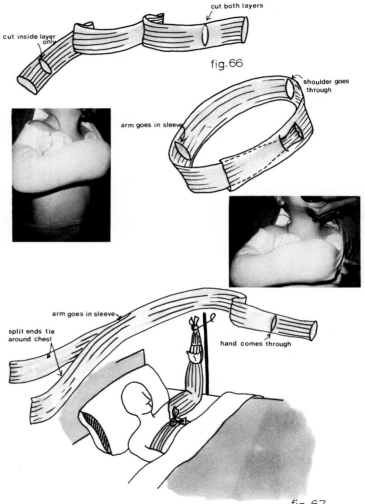

cut both layers
cut inside layer only
fig. 66
shoulder goes through
arm goes in sleeve

arm goes in sleeve
split ends tie around chest
hand comes through

fig. 67

Once the dressings are applied, flat sheets of wet 3-4 inch plaster may be placed on the volar aspect of the hand and forearm, wrapped with KLING and positioned in the SAFE POSITION until hard. Circumferential wide adhesive tape may be applied to immobilize all the dressings. This splint now achieves most of the functions of a circumferential cast, but may be removed easily with a pair of scissors without the need of a cast cutter. (A great convenience for doctors and pediatric patients!)*

These splints may be immobilized in children by a shoulder harness made from stockinettes (Fig. 66). Stockinettes may also be used to strap around a patient's chest to elevate and suspend from an IV pole (Fig. 67).

DURATION OF SPLINTING:

As a general rule, the hand is immobilized following tendon repair as follows:

    Flexor tendons   - 3 weeks
    Extensor tendons   - 5 weeks
    Any tendons at wrist   - 5 weeks
    Mallet, Boutonniere, Recurvatum   - 6 weeks

K-wires for fixation are generally kept in place according to the bones that are being held for healing. In general:

    Phalanges   - 3-4 weeks
    Metacarpals   - 3-4 weeks
    CMC   - 5-6 weeks
    Wrist   - 9 weeks
    SCAPHOID   - 9-12 weeks
    Distal ulna   - 12 weeks

*N.B: This dressing is just as inelastic as a circular cast. Use of Ace bandage not recommended--because of difficulty in judging tightness.

NOTES

NOTES

92 THE HAND BOOK

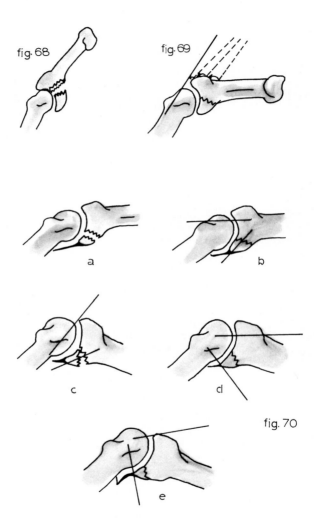

V

Common Injuries and Deformities

PHALANX FRACTURES

Most common fracture injury is hyperextension injury of middle phalanx with volar plate avulsion fracture. If severe, may get dorsal dislocation (Fig. 68).

1. If bone fragment with VOLAR PLATE is not much displaced, may splint alone.
2. To treat CLOSED, pull finger, flex, and "buttress" pin by sliding on dorsum of middle phalanx till it passes into head of proximal phalanx (Fig. 69).
3. If more displaced, need OPEN REDUCTION and pin fixation.
4. IP joint has 4 components: radial and ulnar collateral ligaments, volar plate, and central slip. If you have to sacrifice one with incision, sacrifice ulnar collateral to maintain stability in pinch.
5. May direct-pin (Fig. 70) or buttress-pin fragment.
6. PIN FOR 3-4 WEEKS.

NOTES

NOTES

96   THE HAND BOOK

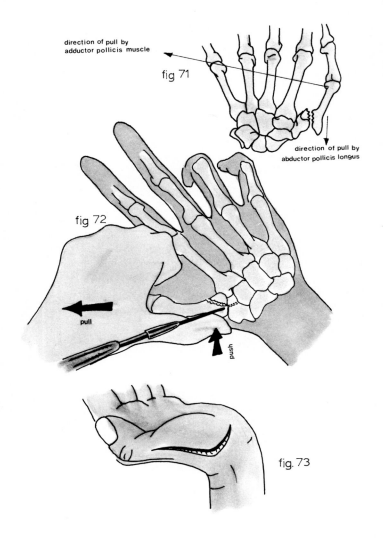

fig 71 — direction of pull by adductor pollicis muscle / direction of pull by abductor pollicis longus

fig 72 — pull / push

fig. 73

METACARPAL FRACTURES

In fractures of the metacarpals, ROTATION deformity is the most unacceptable. To reduce the rotation, the MP joint is flexed to stabilize it, then the finger is reduced in alignment with the other fingers. SLIDING fractures of the metacarpals are shortened and must be lengthened by pulling on the fingers (these often require open reduction). Then the fractured metacarpal may be cross-pinned to the other metacarpals for stability.

SPLINT FOR 3-4 WEEKS.

BENNETT'S FRACTURE

Intra-articular fracture through the base of the first metacarpal. This is LATERALLY displaced through the pull of APL, but medial fragment remains through attachment of collateral ligament (Fig. 71).

- Rx CLOSED - may be reduced by traction but difficult to maintain--may PIN CLOSED into TRAPEZIUM (Fig. 72), K-WIRE X 4 weeks, SPLINT 2 MORE WEEKS.

- OPEN - open reduction through dorsoradial incision on metacarpal (Fig. 73). Align under vision and K-wire across TRAPEZIUM.

BOXER'S FRACTURE

A fracture through the head and/or neck of the fifth metacarpal bone. May accept up to 40° volar angulation. May maintain position with a longitudinal K-wire after reduction, if necessary.

NOTES

NOTES

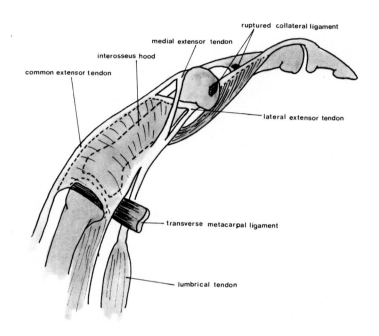

fig. 74

LIGAMENTS

Check for stability with DIGITAL BLOCK after nerve sensation has been checked.

Check ligament for stability by STRESS ROENTGENOGRAM.

Stress IP in EXTENSION, MP in FLEXION.

INCOMPLETE TEAR - Splint in 30-35° flexion for 2-3 WEEKS.

DIP -

Dislocations and ligament tears unusual because skin is snug--would tear.

Rarely unstable even with tears of VP and CL.

SPLINT X 2-3 weeks.

OPEN DISLOCATION - clean, irrigate, reduce.

K-WIRE X 3-4 weeks.

PIP -

For significant displacement must disrupt 2 of 3 components of joint (VP, CL, AL).

<u>Dorsal displacement</u> - tear of VP, AL remains

1. STABLE if VP avulsed fragment<20% of joint surface.

    Rx   CLOSED REDUCTION - repeat x-ray. If unstable, OPEN REDUCTION.

2. UNSTABLE if VP avulsed fragment>50%.

    Rx   OPEN REDUCTION

<u>Lateral displacement</u> - torsion, shearing stress

Radial CL more common than ulnar (6:1).

1. These will present reduced, and MUST STRESS TEST.

    Rx   SPLINT X 3 weeks

2. Some present displaced because fragment of tissue is caught (Fig. 74).

    Rx   OPEN REDUCTION

NOTES

NOTES

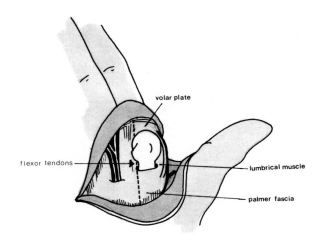
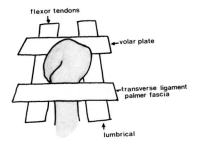

fig. 75

MP -
Usually a dorsal displacement of proximal phalanx over metacarpal head. Metacarpal head is caught in BOX: Lumbrical, volar plate, flexor tendons, and superficial transverse ligament (Fig. 75).

This injury is usually STABLE because of the INTRINSIC muscles. Hyperextension injury causes swelling and ecchymosis.

STRESS TEST WITH MP IN FLEXION.

Rx   Gentle COMPRESSION DRESSING
     SPLINT with MP in slight flexion for 2-3 weeks

Little finger - tear of radial collateral ligament causes ulnar deviation because of pull of ADQ.

Rx   CLOSED   - K-WIRE in OVERCORRECTION X 4 weeks

     OPEN   - if very unstable - REPAIR COLL. LIG.

THUMB -

   1. IP   - like fingers

   2. MP   - Dorsal dislocation - usually treat CLOSED. If displaced with motion: OPEN

           - Lateral dislocation - tear of ulnar collateral ligament (see Gamekeeper's Thumb, p. 109).

   3. CMC - pure dislocation rare; usually a fracture dislocation (see Bennett's fracture, p. 97).

NOTES

NOTES

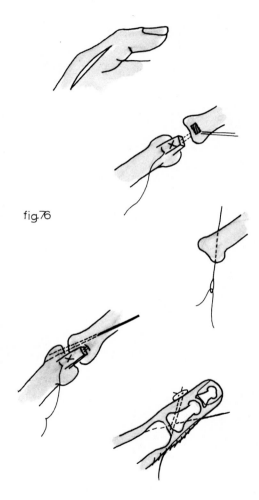

fig. 76

GAMEKEEPER'S THUMB

Laceration of ulnar collateral ligament of thumb MP. Often a ski injury from ski pole.[13,14]

EARLY - Sprain

    Rx   SPLINT X 6-9 weeks, will take 3 months to heal.

LATE - Tear of ligament: instability, pain, lack of pinch.

    Rx   OPEN REPAIR - Tear located between EPL, ADD.P.

1. Incise over ulnar MP joint (Fig. 76).
2. Rongeur piece of bone off proximal phalanx.
3. Bunnell wire through ligament, then drill with Kieth needle through phalanx out radial side - tie over BUTTON.
4. PULLOUT wire through incision.
5. K-WIRE through MP joint.

    PULLOUT WIRE - 3 weeks

    K-WIRE   - 4 weeks

    MOVE   - 5 weeks

LATE, LATE - chronic dislocation. If no degenerative changes: reconstruct an ulnar collateral ligament by:

    Rx   ADDUCTOR ADVANCEMENT to proximal phalanx

LATE, LATE, LATE - degenerative changes. Joint is wiped out.

    Rx   MP FUSION

NOTES

NOTES

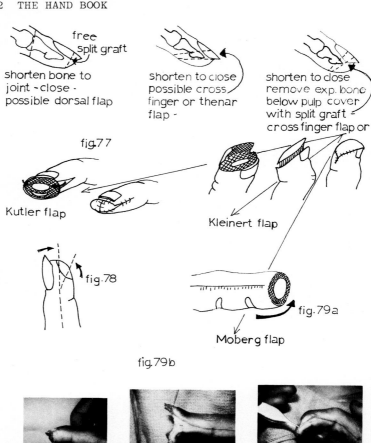

fig. 77 — Kutler flap; Kleinert flap; shorten bone to joint - close - possible dorsal flap; shorten to close possible cross finger or thenar flap -; shorten to close remove exp. bone below pulp cover with split graft = cross finger flap or free split graft

fig. 78

fig. 79a — Moberg flap

fig. 79b

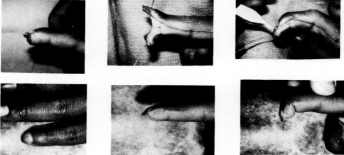

FINGERTIP INJURY

This is one of the most common injuries seen in an emergency room. When the skin loss is significant but a soft tissue pad remains, this may be closed with a skin graft. The best place to get a graft is the inside of the arm, or antecubital crease.

Full thickness skin grafts are better because the defect may be closed primarily and is much less painful than a split graft donor site. DO NOT USE VOLAR WRIST because it may imply suicide attempt when patient goes for job interview.

If the tip is amputated, the wound may be closed with local flaps, e.g., Kutler[15] and Kleinert[16] (Fig. 77). V-Y advancement flaps have the disadvantage that the scar with the 3-corner convergence is at the point of maximal use of a fingertip pad, which is at 45° from the nail (Fig. 78). Volar advancement flaps, as originated by Moberg[17] and described by Snow[18] may be advanced 1-2 cm to cover a defect at the tip (Fig. 79A). This is ideal for the thumb, and should be used with extreme care in the other fingers because the digital vessels are the major blood supply to the dorsal skin of the distal fingers. If used in the fingers, the soft tissue should be left attached as a mesentery to the flap (Fig. 79B).

Remember that the NAIL needs THE DISTAL PHALANX to support its straight growth. IF THE BONE IS AMPUTATED PROXIMAL TO HALF OF THE NAIL, THIS NAIL WILL CLAW, and it would be best to consider removing it.

Also, keep in mind that a pigmented skin graft to the tip of a finger of a black patient would stand out as a "postage stamp". Therefore, would get a more pleasing result by using a full thickness graft of unpigmented palmar skin from ulnar side of hand (at hypothenar area) and close donor site primarily.

NOTES

NOTES

116 THE HAND BOOK

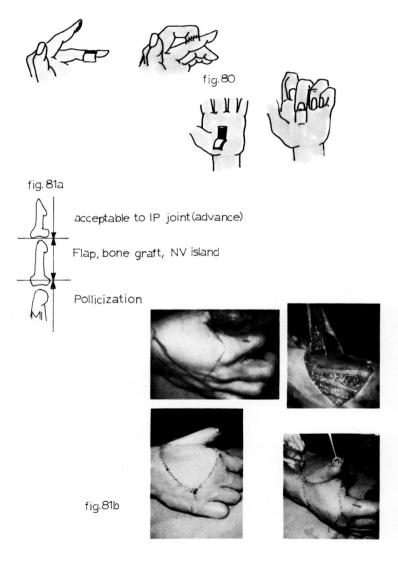

fig. 80

fig. 81a

acceptable to IP joint (advance)

Flap, bone graft, NV island

Pollicization

fig. 81b

For loss of larger areas of the fingertip or of the pad, cross-finger flaps and palmar flaps may be used for resurfacing (Fig. 80).

Thumb -

Moberg volar advancements are ideal for loss up to 1.5 - 2.0 cm of the thumb tip. The thumb is functional up to the articulation (Fig. 81A). If the amputation is up to the base of the proximal phalanx, a bone graft and NV island pedicle is ideal. If the base of the proximal phalanx is lost, pollicization should be considered.

Ring Avulsion -

Depending on age, occupation and the particular needs of each patient, repairs would consist of microvascular repairs, flaps, grafts, and NV island transfers.[19]

Sometimes a neurocutaneous flap from the dorsum of the hand may be used to resurface a ring-avulsion of a thumb, as long as the radial nerve is left intact in this flap (Fig. 81B).

NOTES

NOTES

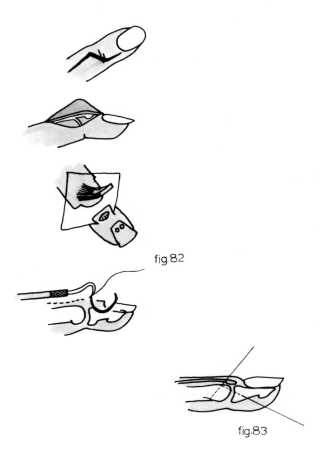

fig. 82

fig. 83

MALLET FINGER

This is a deformity that is a result of disruption of the extensor tendon from its insertion at the base of the distal phalanx. An x-ray of the finger may show a fragment of the bony insertion avulsed off the phalanx.

1. If seen immediately or EARLY, splint in hyperextension with plaster or aluminum splint and/or K-wire X <u>6 WEEKS.</u>

   If seen LATE, splint for 3-4 weeks, if improved, splint another 3-4 weeks.

   If seen VERY LATE, or no improvement with splinting; OPERATE:

2. Through a dorsal Z-incision (Fig. 82), identify and dissect insertion of central slip (may have healed long).

3. Chisel out bone at insertion of phalanx, and drill two holes from nail bed into defect in phalanx (large enough for cleft palate needle).

4. K-wire DIP joint in extension.

5. Pull Bunnell wire through extensor tendon into eye of needle in bony defect, and tie over button on nail.

6. If fragment of bone on end of tendon, may buttress pin (Fig. 83).

7. SPLINT X 6 WEEKS.

May try TENODERMODESIS:[20]

1. Anesthetize finger by block, and apply tourniquet.

2. Excise ellipse of skin and tendinous scar tissue down to bone over dorsum of DIP joint.

3. SPLINT DIP JOINT IN EXTENSION X 6 WEEKS.

NOTES

NOTES

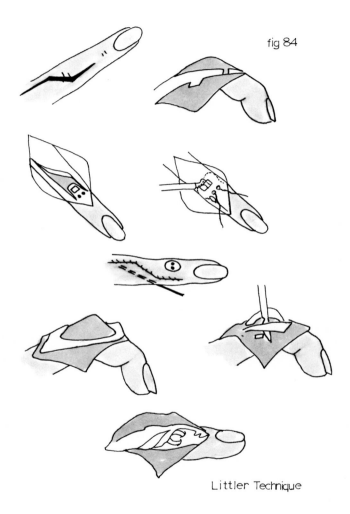

fig 84

Littler Technique

BOUTONNIERE DEFORMITY

This is a result of disruption of the CENTRAL SLIP of the extensor tendon from the insertion at the base of the middle phalanx. Initially, this injury can be missed until the forces of action move the LATERAL BANDS volarly below the axis of rotation of the joint. At that point the lateral bands will FLEX the PIP and EXTEND the DIP to result in the deformity.

The BOUTONNIERE TEST: passive extension of the PIP hyper-extends the DIP joint (see p. 65).

1. Passively extend the PIP joint (with a JOINT JACK) to stretch out the lateral bands and allow dorsal replacement.

2. Through a dorsal Z-incision (Fig. 84), cut the volarly displaced ULNAR lateral band and retinacular ligament. Preserve the lateral band on the radial side which contains the tendinous extensions of the Lumbrical muscle.

3. Chisel out bone at area of insertion at base of middle phalanx, and drill 2 holes retrograde into this defect.

4. K-wire PIP joint in extension.

5. Using cleft palate needles passed retrograde into defect, pass Bunnell wire out and tie over button.

6. K-WIRE X 4 weeks, SPLINT X 6 WEEKS.

NOTES

NOTES

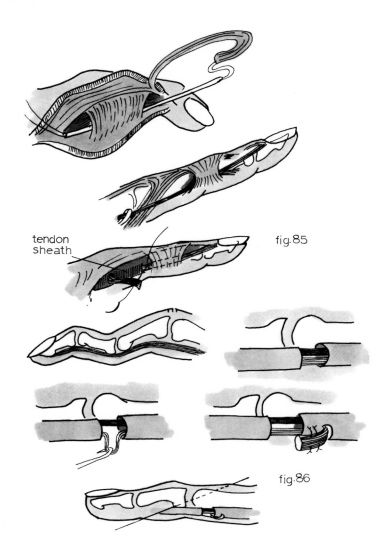

tendon sheath

fig. 85

fig. 86

RECURVATUM DEFORMITY
(SWAN NECK DEFORMITY)

This deformity is due to imbalance of tendon forces from trauma, degenerative disease, or congenital.

A. <u>LITTLER TENODESIS</u> (Fig. 85): <u>ulnar lateral band</u> under Cleland's ligament to flexor sheath to act as an oblique retinacular ligament.

B. <u>FLEXOR TENODESIS</u> (Fig. 86): <u>one slip of FDS</u> to flexor sheath to prevent hyperextension.

1. Mid axial incision on non-dominant side.

2. Incise flexor sheath NO MORE DISTAL THAN MID-PROXIMAL JOINT--flex wrist, get FDS, cut ONE SLIP proximally.

3. K-WIRE IP joint in neutral extension.

4. Route distal end of cut sublimis through flexor sheath and suture to sheath and itself.

5. DO NOT CUT SHEATH DISTAL THAN MID-PROXIMAL JOINT OR IT MAY SLIP OUT AND <u>HYPEREXTEND.</u>

6. SPLINT 3-4 WEEKS.

NOTES

NOTES

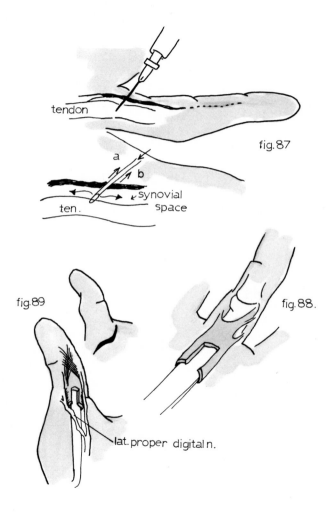

NON-SPECIFIC TENOSYNOVITIS
(STENOSING TENOSYNOVITIS)

A. TRIGGER FINGER: patients with STS (stenosing tenosynovitis) wake up in the morning with finger caught in flexion, click, and pain on extension. This loosens during day.

   Non-specific inflammation of flexor tendon sheath heals by scar and stenosis of tunnel opening. The tendon is soft and "doughy" and during sleep at night, edema causes swelling of tendon on both sides of stenosis (dumbell-shape), with fingers flexed (normal position during sleep). In morning, finger "caught" in this position. Continued use during day "squeezes" fluid from tendon, and lessens trigger phenomenon.

   1. May inject with steroid into sheath (Fig. 87). After needle is introduced in tendon, gentle pressure on plunger while withdrawing needle will inject the medication when the needle is properly in sheath space.

   2. "Rule of 3's": Xylocaine may relieve for 3 hours. Symptoms may flare up right after injection and may take 3 days for steroids to take effect. Steroid effect lasts 3 weeks. May inject 3 times, if symptoms return, operate.

   3. A transverse incision over the flexor sheath at MP joint and excision of a segment of the proximal stenosed portion will release the tendon (Fig. 88).

   4. Beware of the digital nerves, particularly in the thumb where they lie in a more volar position (Fig. 89).

B. ACUTE FCU TENOSYNOVITIS: severe pain, patient crying, will not let hand be examined. Wrist red, puffy.
   Dx: infection or gout
   Rx: steroid injection and plaster splint

   If very severe, give prednisone and phenylbutazone, splint, and give codiene for 24 hours to slow down, then inject steroids.

NOTES

NOTES

136 THE HAND BOOK

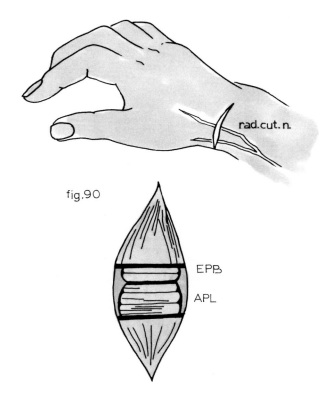

fig.90

DEQUERVAIN'S DISEASE

Tenosynovitis of first dorsal compartment of extensor tendons. Diagnosis made by thickness and tenderness over this compartment, and by Finkelstein Test (see p. 65). Differential diagnosis is DEGENERATIVE ARTHRITIS of CMC JOINT.

1. Probably SHOULD NOT BE INJECTED with steroids because it will be very painful, and also because they recur and invariably need surgery.

2. Transverse incision over the radial aspect of wrist to release the tendon sheaths (Fig. 90).

3. APL usually has 2-3 SLIPS. First 1 or 2 go to APB muscle and 1 goes to thumb metacarpal.

4. EPB may also have 2-3 SLIPS. Make sure all the tunnel sheaths are open.

5. Make sure the most dorsally cut sheath has the EPB.

6. Watch for branches of the CUTANEOUS BRANCH of RADIAL NERVE. It will cause significant annoying neuroma when cut. If cut, repair to itself.

NOTES

NOTES

140   THE HAND BOOK

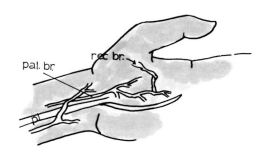

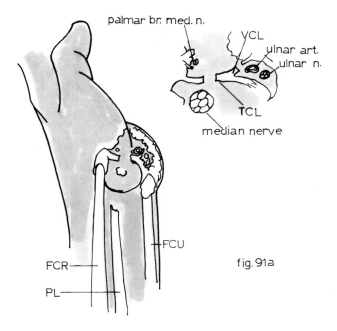

fig. 91a

CARPAL TUNNEL SYNDROME

Compression of the median nerve secondary to scarring and thickening of the transverse carpal ligament. Has 3 components:[21]

1. Night PAIN in 3-4 hours. May radiate to elbow or shoulder.
2. MORNING STIFFNESS due to "catching" of flexor tendons.
3. DAYTIME NUMBNESS in wrist-flexed position, such as driving, typing, sewing, telephone.

Tests:

Phalen's, Tinel's (see p. 69).

Two-point sensation.

Thenar muscle weakness - resistance to opposed thumb (Fig. 47B, p. 64).

Hypothenar weakness - resistance to abducted little finger.

1. Probably should not inject because they invariably recur and need surgery.
2. Incision is vertical up to but not proximal to wrist crease (Fig. 91A). Avoid palm. cut. br. of median nerve which is radial to PL.
3. Expose superficial palmar arterial arch and follow into ulnar canal (Guyon) to protect this nerve. Stay on radial side of ulnar canal to protect ulnar nerve.
4. Open roof of Guyon's canal (volar carpal ligament).
5. Find FDS to ring finger and follow into carpal tunnel, cutting on most ulnar aspect (protects palm. cut. br. which is radial to PL).
6. Expose motor branch of median nerve (see if around or through TCL). Be aware of anatomical variations.[22,23]

NOTES

NOTES

144 THE HAND BOOK

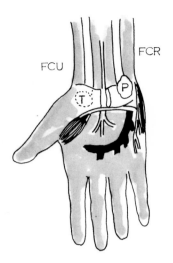

fig.91b

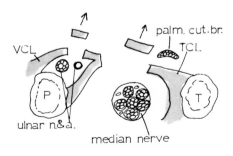

7. Cut two layers of "sandwich" of TCL and thenar muscles to EXCISE SEGMENT of TCL (Fig. 91B).

8. Perform neurolysis by excising thickened epineureum of median nerve.

9. SPLINT X 1 WEEK.

   In patients with diabetes or Dupuytren's disease, move early (3-4 days) to prevent post-op stiffness. In bilateral cases, repair one hand at a time, 10 days apart.

   May get flare-up of STS post-operatively.[21]

NOTES

NOTES

148   THE HAND BOOK

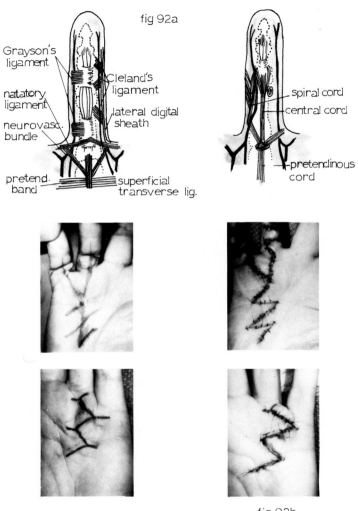

fig 92a

fig. 92b

DUPUYTREN'S CONTRACTURE

A proliferative disorder of the PALMAR FASCIA which may result in contractures of various components and result in representative deformities. Is believed to be associated with alcoholism, with cirrhosis, patients with seizure disorders on barbiturates, with tuberculosis, with Peyronie's disease. Contracture has been identified with the presence of myofibroblasts in the contracted nodules.

KNUCKLE PADS often appear earlier than the contracture. These should not be surgically treated unless very painful or involve the hood mechanism or joint underneath.

SKIN NODULE is first presentation, but this may not progress to contracture for a long time. This should not be treated unless it is painful (may have trapped cutaneous nerve). When only NODULES are present, if hands are placed at REST because of trauma or medical disability, this will lead rapidly to CONTRACTURE.

CONTRACTURES may involve any combination of the components of the palmar and digital fascias: pretendinous band, spiral band, natatory ligament, lateral digital sheet, Grayson's ligament, Cleland's ligament (Fig. 92A).[24]

Surgical treatment of the contracture requires lengthening of the skin that has been contracted. This may be accomplished by a limited palmar fasciectomy with removal of the involved contracted components made through skin incisions that employ lengthening. Multiple Z-plasties on a midline incision allow for maximal lengthening but may result in compromized flaps. Multiple V-Y advancements of skin of fingers on palm allow for lengthening of skin without undermining or compromise to the flaps (Fig. 92B).

In severe flexion deformities in patients who cannot tolerate prolonged immobilization (e.g., elderly), the "open palm" technique may be employed.[25,26] In this technique the wound is allowed to heal by secondary intention to close the transverse incision used to perform the fasciectomy.

NOTES

NOTES

fig. 93

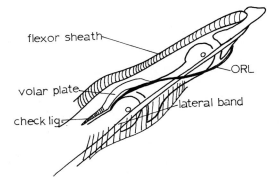

DIGITAL FLEXION CONTRACTURE

May be due to Dupuytren's, traumatic crush, or scar contracture (Fig. 93).

May first try to correct by JOINT JACK to straighten out scarred structures or pull adhesions.

1. In Dupuytren's first excise palmar fascia and vertical septa.
2. Otherwise, use Bruner or mid-axial incision. If mid-axial, will need bilateral incisions.
3. Cut contracted FLEXOR SHEATH.
4. Cut TRANSVERSE LAMINA of LANDSMEER (TLL) on joint hood to release lateral bands.
5. If LATERAL BANDS are volar and contracted, cut these--may need to do transfer of lateral bands as in Boutonniere repair.
6. Cut the OBLIQUE RETINACULAR LIGAMENT.
7. May need to cut the CHECK-REIN LIGAMENTS--cut on periosteum proximal to joint--beware of vessels (see p. 57).
8. If necessary, cut the VOLAR PLATE, may need to re-establish normal pouch beneath proximal head (see Fig. 96, p. 160).
9. Do these stepwise, extending the joint passively-- once joint can be extended need not release any other structures in sequence.
10. K-WIRE X 3-4 weeks. SPLINT 6 weeks.

NOTES

NOTES

NOTES

In Dupuytren's flexion contracture, do extensor tenolysis of lateral bands. May need added power--

- ring finger ulnar lateral band to little finger radial lateral band
- EIP transfer
- ADQ transfer
    1. Cut ulnar lateral band to little finger.
    2. Cut ADQ from phalanx, weave into ulnar lateral band.
    3. Acts as DYNAMIC TENODESIS:

        MP extension tightens ADQ, extends DIP.

        MP flexion contracts ADQ, extends DIP.

NOTES

NOTES

160 THE HAND BOOK

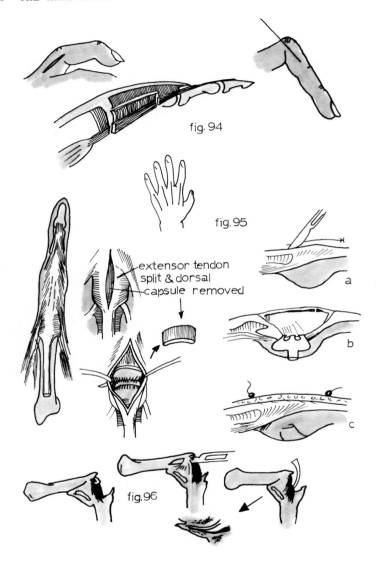

fig. 94

fig. 95

extensor tendon split & dorsal capsule removed

a

b

c

fig. 96

COMMON INJURIES   161

EXTENSION CONTRACTURE

Extrinsic extensors and intrinsic muscles extend the interphalangeal joints. They extend the joints with the coordinating action of the volar retinacular ligament and oblique retinacular ligament of Landsmeer, (which acts as a dynamic tenodesis).

Function is little impaired if either one or the other component of the conjoined extensor mechanism is removed just distal to the MP joint. Adhesions or contractures of the head result in extension contracture.

LITTLER RELEASE

1. Bilateral mid-lateral incisions are made to expose the extensor hood (Fig. 94) (or dorsal midline incision).

2. Cut a small triangle out to release. If inadequate, do the other side. If still inadequate, take larger triangle.

3. Passively flex MP joint until released.

4. K-WIRE MP joint in 90° flexion X 1 week.

CAPSULECTOMY

If there is no flexion of metacarpophalangeal joint, but intrinsically normal joint, motion may be regained by a capsulectomy.

1. The extensor tendons may be exposed by longitudinal incisions (Fig. 95).

2. The tendon is split and the extensor hood is retracted to expose the collateral ligaments.

3. The dorsal portion of the capsule is excised.

4. A portion of the collateral ligaments may need to be excised.

5. A normal pouch of the volar plate may need to be re-established underneath the metacarpal hood (Fig. 96).

NOTES

NOTES

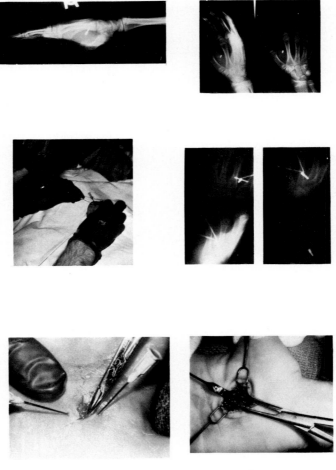

fig.97

FOREIGN BODIES

When a foreign body is seen radiographically, the hand may be blocked with local anesthesia and the foreign body may be located stereotactically with a fluoroscope and image intensifer.[27] Under aseptic control, two 19 gauge needles may be placed from opposite directions to pinpoint the foreign body (Fig. 97).

The 3-dimensional stereoscopic effect is obtained by the patient moving the hand upon command while the surgeon views the television screen of the image intensifier. The x-ray beam is shut off for each manipulation of the localizing needle for a new position, so that there is no radiation exposure to the hands of the operator.

Once the foreign body is localized, a small incision is made between the two needles down to the converging tips to locate the foreign body.

Beware of any extensive probing! Include tourniquet control if foreign body located deep.

NOTES

NOTES

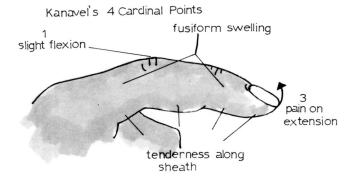

fig. 98

INFECTIONS

In general, infections of the hand or fingers need to be treated conservatively. If there is associated systemic effects (temperature over 101°, elevated white blood count) or cellulitis, or lymphangitis, the patient should be admitted to the hospital for intravenous antibiotic therapy.

The hand should be immobilized in the SAFE POSITION, and elevated. Wet dressings are not necessary. The natural defense mechanism of the skin against bacteria is by virtue of the dry keratin layer which causes dessication of bacteria, and fatty acids secreted by the glands which kill streptococci. Since a cellulitis already has maximal capillary dilatation, hot soaks will not increase the circulation, but will wash away the fatty acids and will soften the keratin and break this natural barrier.

The only logical use of wet dressings would be to prevent drying and crusting of purulent discharges after an abscess has been drained.

Therefore, an infection of a finger or hand needs to be immobilized and elevated. The antibiotics will either resolve the infection or localize it to a smaller collection, and if an abscess forms, this may then be incised and drained.

SUPPURATIVE TENOSYNOVITIS is an infection that has involved the flexor sheath of the finger. The diagnosis may be made according to KANAVEL'S 4 CARDINAL POINTS (Fig. 98):[28,29]

1. Slight flexion of finger

2. Fusiform swelling of finger

3. Pain on extension (passive or active)

4. Tenderness along tendon sheath (into palm)

These may require surgical drainage when purulent collection is found. Sometimes, small plastic catheters may be placed into the infected space to provide continuous irrigation with antibiotic solutions.[30]

NOTES

NOTES

172 THE HAND BOOK

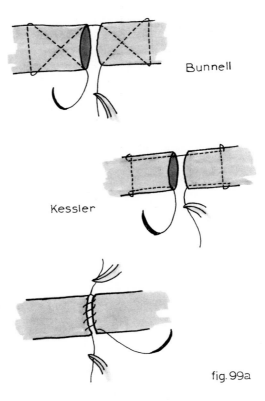

Bunnell

Kessler

fig. 99a

VI

Flexor Tendon Injuries

LACERATION

GENERAL PRINCIPLES:

The treatment for laceration of a flexor tendon is dependent on the level of the laceration, dominance of the hand, age, and occupation of the patient.

"NO MAN'S LAND" is located between the DISTAL PALMAR CREASE and the DIP flexion crease, and represents the portion of the tendon within the FIBRO OSSEOUS CANAL where scars and adhesions form readily.

DISTAL:
    Less than 1.5 cm from insertion of FDP
    Rx: advancement and insertion of FDP

MIDDLE (within fibro osseous canal):

    More than 1.5 cm

    FDS INTACT:

    1. Nothing (laborer, older patient-ring or little finger*)
    2. DIP tenodesis
    3. Primary repair
    4. Primary or delayed tendon graft

    FDS CUT:

    1. Primary repair of FDP only
    2. Primary repair of both FDP and FDS
    3. Primary or delayed tendon graft

*RING AND LITTLE FINGERS:
    Laborers do not need the ring and little finger flexion at the DIP joint and may do very well with a tenodesis which allows them to return to work earlier. However, laborers need FDS in the ring and little fingers because these give them POWER GRASP (e.g., hammer, screwdriver). The FDS to the little finger is thin, narrow, and not strong. Therefore, if FDP is cut, repair is needed for strength and grasp.

NOTES

NOTES

176 THE HAND BOOK

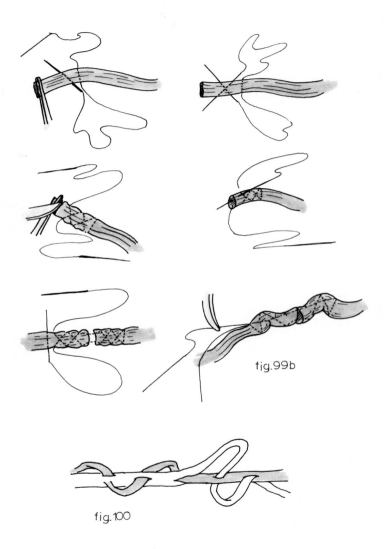

fig. 99b

fig. 100

PROXIMAL (proximal to distal palmar crease):

    FDS ONLY:
1. Nothing
2. Primary repair in musician (requires independent PIP flexion)

    BOTH TENDONS:
1. Primary repair of FDP only
2. Primary repair of FDP and FDS

INDICATIONS FOR PRIMARY REPAIR
(or delayed primary in 7-14 days)
1. Children (do very well)
2. Cut in palm, wrist, forearm
3. Cut in fibro osseous tunnel (according to Kleinert)

A purposefully delayed primary repair may be justified in certain circumstances.[31]

INDICATIONS FOR HUNTER ROD
(and delayed free graft)
1. Scarring in wound
2. Fractures (holds tunnel open until fractures heal)
3. Both digital nerves cut (holds till they have sensation)
4. Delayed free graft planned (keeps pulleys patent till return)

BUNNEL STITCH--is the most frequently used technique for flexor tendon suture. A non-absorbable 4-0 suture is passed through the tendon, criss-crossed twice, coming out the cut end (Fig. 99A). This may be passed through the bone for direct insertion, or it may be sutured to the opposite cut end of tendon. If stainless steel wire is used, a pull-out wire may be passed to remove this later (see p. 181). The tendon may also be sutured by a modified BUNNEL stitch (one - X), or a KESSLER stitch. The tendon repair may be completed with a running 6-0 suture closing raw edges.

TENDON WEAVE

    The tendon may be woven through another tendon at 90° angles, to cinch up the proper tension (Fig. 100). With the proper tension maintained, the corners of the weave are sutured with 3-corner stitches with fine sutures. The ends of the tendons are then cut flush with the walls that they penetrate, and these cut ends retract to be buried within the tendon interior and sutured over.

NOTES

NOTES

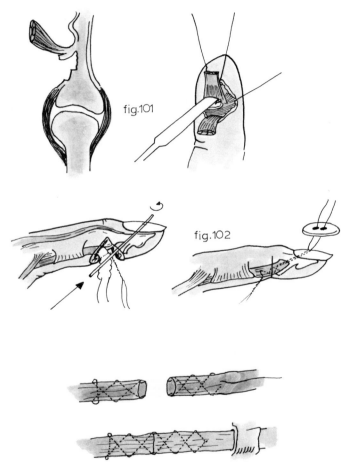

fig.101

fig.102

fig.103

REPAIR OF LACERATED FLEXOR TENDON

ADVANCEMENT:

1. Incise on non-dominant side of finger to expose the sheath and distal tendon end.
2. Prepare bed at distal phalanx by elevating periosteum (Fig. 101).
3. Drill hole through this bed and out the dorsum of nail. Pass small wire loop retrograde (Fig. 102).
4. Pass Bunnel suture through end of proximal tendon (with pull-out wire). Then pass through wire loop and out the distal phalanx (Fig. 102).
5. Tie over button. Pass pull-out wire through skin.
6. SPLINT X 3 WEEKS.

END-TO-END REPAIR

1. Pass Bunnel suture through cut end of proximal tendon with non-absorbable suture (e.g. nylon or wire).
2. Cinch up and tie (Fig. 102).
3. SPLINT X 3 WEEKS.

NOTES

NOTES

184 THE HAND BOOK

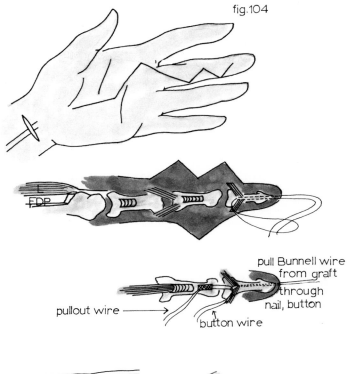

fig.104

pull Bunnell wire
from graft
through
nail, button

pullout wire

button wire

fig 105

FREE TENDON GRAFT

Wait until there is full free range of motion. If a Hunter Rod has been placed, make proximal tunnel into wrist so rod has free motion on flexion.[32] Wait till full passive range of motion before inserting free tendon graft (4-6 weeks). Optimal time is 3-4 weeks if free motion exits because pseudosheath is thinnest at this time, and increases thickness thereafter.[33]

1. Use volar incision to expose area (Bruner or Littler).

2. Excise flexor sheath except for pulleys.

3. If pulleys are not present--reconstruct them (see p.189).

4. Prepare bed in distal joint--drill hole through nail, pass loop of wire from nail into bed (Fig. 104).

5. Expose profundis tendon, lumbrical and NV bundle in palm through fascial compartment on ulnar side of lacerated tendon.

6. For tendon graft, if SUBLIMIS is cut use this. Otherwise, PL, RING SUBLIMIS (see sublimis tenodesis p.193), PLANTARIS, EIP, EDQ, TOE EXTENSOR.

7. Pass BUNNEL WIRE through end of graft (with pull-out wire) and pass from palm through pulleys and through distal phalanx drill hole by means of wire loop and tie-over button. Pass pull-out wire through skin on lateral side of finger.

8. Suture finger incisions closed.

9. Weave proximal end of graft through FDP near origin of Lumbrical (Fig. 105).

10. Pull tendon and graft till proper tension brings finger in line with other flexed fingers and suture woven corners.

11. SPLINT IN FLEXION X 3 WEEKS.
    (Keep wrist 15° - 20° short of full flexion)

NOTES

NOTES

188   THE HAND BOOK

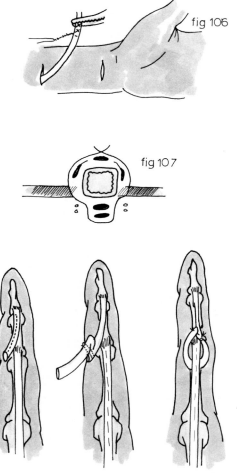

fig 106

fig 107

fig 108

PALMARIS LONGUS GRAFT

Two incisions are made--one at the wrist flexion crease to get insertion on palmar fascia, and one proximally when the distal end of tendon is pulled on tension (Fig. 106).

DO NOT HANDLE TENDON WITH FINGERS OR CRUSHING INSTRUMENTS.

Place clamp on distal end of tendon--pass Bunnel wire and pull-out wire. Cut distal crushed tendon at free end of Bunnel suture. Then cut proximal end of tendon and transfer to finger.

FLEXOR PULLEYS

If a flexor pulley is needed, a segment of tendon can be used from a cut FDS, or a segment may be harvested from PL.

This free segment of tendon is passed around the finger circumferentially (around the dorsum between the extensor tendons and the skin). The ends are woven and sutured. Then the tendon pulley is rotated so that the woven ends are located dorsally (Fig. 107).

PROFUNDIS TENODESIS

Do tenodesis in laborers only. Do not settle for tenodesis in little finger as the FDS is very thin and tenuous.

A simple technique to perform a tenodesis of the DIP joint[34] is to take the distal cut end of the FDP, incise and lengthen this tendon, slip it around the insertion of the FDS and suture it to itself (Fig. 108). This should be pulled up snugly to produce flexion at the DIP joint at surgery.

NOTES

NOTES

192 THE HAND BOOK

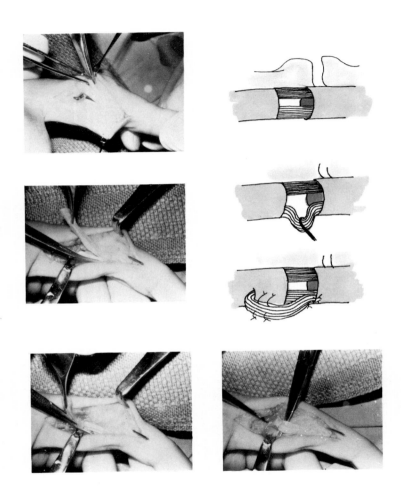

fig.109

SUBLIMIS TENODESIS

If the FDS is cut but has no restraint, leave it alone for it forms:

--an excellent bed for overlying graft with minimal scar

--protection of check-rein ligaments at base of volar plate

--protection to the vincular blood supply and the transverse branch of the digital arteries (see p. 57).

To prevent recurvatum deformity (hyperextension of PIP joint) one slip may be used to suture to flexor sheath to form tenodesis (Fig. 109). If ring sublimis tendon is used for grafts, suture one slip of tendon to flexor sheath to prevent hyperextension of joint.

FLEXOR TENDON RUPTURE

Sudden rupture of the flexor tendons are not as common as in the extensor tendons. When they do occur, they are usually of the PROFUNDUS at its insertion at the distal phalanx. Most commonly, it occurs in young men as a result of athletic injuries. Most commonly the RING FINGER[35] gets caught in opponent's football jersey or pants as the other fingers slip out while opponent continues to pull away. Treatment is the same as lacerations: advancement, repair, or free tendon graft.

PARTIAL LACERATION OF TENDON

Repair of partial lacerations of flexor tendons may not be uniformly accepted. Although repair of partial lacerations in chickens decreases both tensile strength and glide.[36] Kleinert feels all should be repaired.[37] Nevertheless, a series of 17 consecutive partial lacerations treated without suturing and with early motion resulted in excellent function.[38] This concept should be considered as a possible method of treatment.

NOTES

NOTES

196 THE HAND BOOK

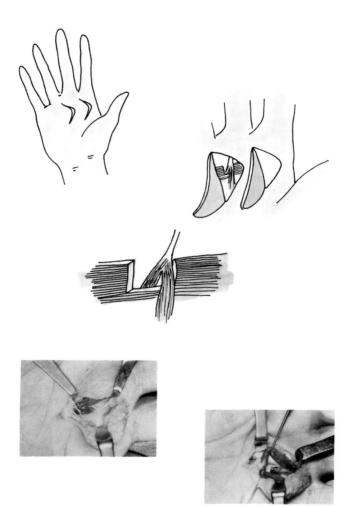

fig.110

VII

Extensor Tendon Injuries

Lacerations of the extensor tendon over the dorsum of the finger may be treated as described previously under Mallet finger, Boutonniere deformity, and Recurvatum deformity.

EXTENSOR OVER DORSUM OF HAND

Lacerations of this area may be repaired by a modified Bunnell suture (see flexor tendon repair) tied over buttons (Fig. 111). ALTERNATIVELY, it may be sutured with 2 or 3 "Figure of 8" sutures through the skin (Fig. 112); then the lacerations may be closed separately. The hand should be splinted with the wrist in dorsiflexion, MP's at 30° and IP's at 180° for 4 WEEKS.

EXTENSOR POLLICIS LONGUS

When the EPL is lacerated in a clean wound, it should be repaired primarily. If this repair heals with scarring, or if the patient presents late with scarred wound, it is better to transfer the EIP to the EPL.

1. A dorsal incision over the index MP will expose the EIP on the ulnar side of the EDC and is cut.

2. A dorsal incision over the scaphoid bone will reveal the EIP here, may be pulled back and tunneled under the skin to re-route to the ulnar side of the thumb MP. Here the tendons are repaired by the weave technique (see flexor tendon repair).

LOSS OF EXTENSOR DIGITORUM COMMUNIS

An avulsion injury that removes segments of the EDC may be repaired by a secondary graft. Usually the wound is dirty.

1. Clean wound and close laceration or allow flap to heal in secondarily.

2. Once the wound has healed, the area can be exposed and proximal and distal ends are repaired with a graft (PL, FDS, Plantaris, Ext. Hallucis) (Fig. 113A).

NOTES

NOTES

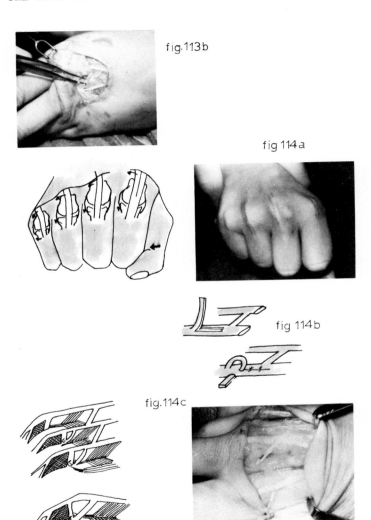

fig. 113b

fig 114a

fig 114b

fig. 114c

SILASTIC SHEETS

When extensive scarring is present in the wound, a small piece of thin silastic sheet may be placed on the wound bed beneath the repair or distance of graft to prevent adhesions to the bed (Fig. 113B). Adhesion of the tendon to the overlying skin usually does not limit motion as the skin can glide with the excision.

EXTENSOR TENDON DISLOCATION

The dislocation of the extensor tendon over the dorsum of the MP joint (Fig. 114A) may be secondary to:

1. congenital
2. degenerative (rheumatoid)
3. traumatic - flexion/torsion stress causing radial tear which is more commonly seen in the long and ring fingers[40,41,42]

Procedure:

1. Slip of extensor tendon is stripped in the middle and passed radially (Fig. 114B). The tendon is then closed with fine sutures to prevent splitting of tendon.

2. This slip is then sutured to the sagittal band on the radial aspect through a weave (Fig. 114C). This also pulls the sagittal band proximally on extension.

3. If a discreet sagittal band is not present, may suture this slip to the transverse metacarpal ligament, or to the adjacent extensor tendons (if not subluxed).

NOTES

NOTES

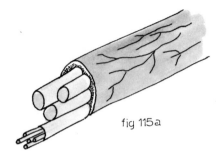

fig 115a

VIII

Nerve Repairs

Anatomy:

Individual axons are enclosed by an ENDONEURIUM. These units bunched together form FASCICULI (or funiculi) and are covered by PERINERIUM. These fasciculi bunched together form the nerve which is covered by an EPINEURIUM (Fig. 115).

Injury:

The quality of sensation decreases with time between the injury and the repair of the nerve, but is said to never fall to zero. However, when the time interval for repair of a motor nerve is long, or the distance necessary for axons to travel from a very proximal injury is long, the failure is due to muscle atrophy and scarring that takes place in the interval.

Lacerated nerves should be repaired primarily, or as soon thereafter as possible. Delayed primary repairs (4-10 days) do as well, but repairs delayed beyond 2-3 weeks do not do well because of technical problems. If a repair needs to be delayed, it is best to tag the ends with non-absorbable sutures for easy identification later on.

Millesi has demonstrated excellent results with inter-fascicular repair and nerve-grafts, and clearly showed that the most important factor influencing the result of nerve repairs is tension.[43] Although microsurgical techniques have facilitated the repair of nerves by interfascicular stitching,[44] the controversy remains.[45,46]

Sensory recovery has been shown to be further facilitated by a sensory re-education program to improve the results from nerve repairs.[47] This technique employs the use of nuts and bolts, buttons and objects with gradually decreasing sizes to educate sensory perceptions.

NOTES

NOTES

212 THE HAND BOOK

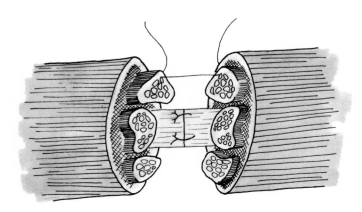

fig 115b

DIGITAL NERVES

Laceration of digital nerves distal to the DIP flexion crease may be left unrepaired; beyond this the nerve begins to branch extensively. Sensation usually returns (though not necessarily to normal) from regenerating axons or from the digital nerve from the opposite side of the finger. Lacerations proximal to the DIP flexion crease may be repaired by means of loop or microscopic magnification with 8-0 to 10-0 sutures in the epineurium.

SPLINT X 3-4 WEEKS.

Larger nerves at the wrist may be repaired with interfascicular sutures under magnification (Fig. 115B). These hands should be immobilized in plaster (wrist flexion for median and ulnar nerves) following the repair.

SPLINT X 3-4 WEEKS.

NOTES

NOTES

220 THE HAND BOOK

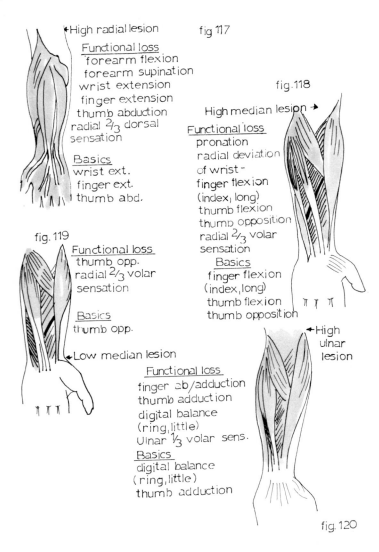

←High radial lesion   fig 117
Functional loss
 forearm flexion
 forearm supination
 wrist extension
 finger extension
 thumb abduction
 radial 2/3 dorsal
 sensation
Basics
 wrist ext.
 finger ext.
 thumb abd.

fig. 118
High median lesion →
Functional loss
 pronation
 radial deviation
 of wrist -
 finger flexion
 (index, long)
 thumb flexion
 thumb opposition
 radial 2/3 volar
 sensation
Basics
 finger flexion
 (index, long)
 thumb flexion
 thumb opposition

fig. 119
Functional loss
 thumb opp.
 radial 2/3 volar
 sensation
Basics
 thumb opp.

←Low median lesion
Functional loss
 finger ab/adduction
 thumb adduction
 digital balance
 (ring, little)
 Ulnar 1/3 volar sens.
Basics
 digital balance
 (ring, little)
 thumb adduction

←High ulnar lesion

fig. 120

IX

Tendon Transfers

    To determine appropriate tendon transfers in nerve palsies, determine loss of muscles, and available muscles. Different combinations may be employed to substitute function.

RADIAL NERVE - (Fig. 117)

    Lose - wrist extension
           finger extension
           thumb extension

HIGH MEDIAN NERVE - (Fig. 118)

    Lose - wrist flexion
           finger flexion (all FDS and FDP to index and
                      long finger lost)

           thumb flexion
           thumb opposition

LOW MEDIAN NERVE - (Fig. 119)

    Lose - thumb opposition

ULNAR NERVE - (Fig. 120)

    Lose - thumb adduction
           intrinsic function (clawing) (low ulnar)
           finger flexion (FDP to ring & little)
                    (high ulnar-no claw to ring and
                    little because lose FDP also)

NOTES

NOTES

224 THE HAND BOOK

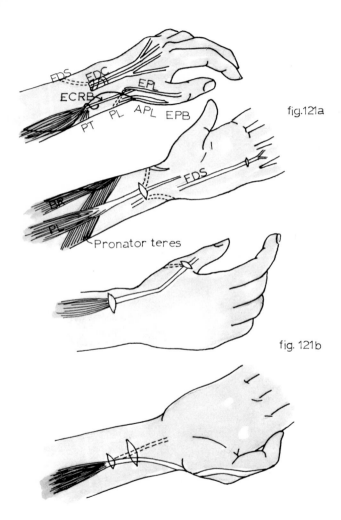

fig. 121a

fig. 121b

RADIAL (Fig. 121A)

WRIST EXTENSION

- * Pronator Teres ⟶ ECRB
-   FCU ⟶ ECRB
-   FCU ⟶ ECU
-   APL ⟶ ECRB

FINGER EXTENSION

- * Ring Sublimis ⟶ EDC
  (sublimis has greater excursion, more power)
-   Pronator Teres ⟶ EDC
-   FCU ⟶ EDC
  (less excursion, less power)

(In cerebral palsy when radial nerve to BR may be intact, use BR ⟶ EDC, more power.)

THUMB EXTENSION

-   Palm long ⟶ EPL (Fig. 121B)
- * EIP ⟶ EPL

NOTES

NOTES

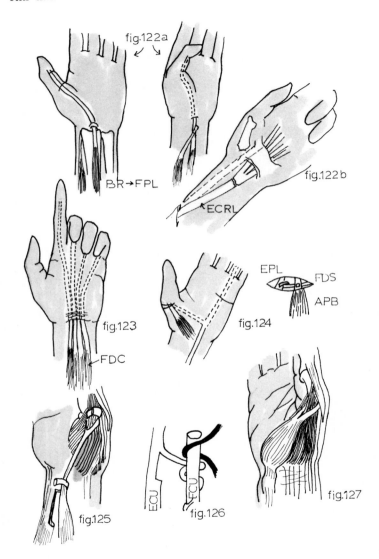

MEDIAN

THUMB FLEXION

BR ⟶ FPL (Fig. 122A)

FINGER FLEXION

BR ⟶ Tethered FDP
ECRL ⟶ Tethered FDP (Fig. 122B)

(High median - tether FDP together - FDP to ring and little are functional through ulnar nerve (Fig. 123).

THUMB OPPOSITION (Fig. 124)

* Ring Sublimis ⟶ APB & EPL

ADQ ⟶ APB & OP

EDQ ⟶ APB & OP

ECU ⟶ APB & EPL

ECRL ⟶ APB & EPL

Palm Long ⟶ APB & EPL

If ABDUCTION is present in thumb (low median nerve) may use a distal pulley for opposition. If ABDUCTION is lost (high median) use proximal pulley around FCU to abduct as well as oppose.

Distal pulley (Royle-Thompson)[48,49] - Palmar fascia as pulley. (Brand) Septa in thenar skin around hook of hamate, pisiform.

Proximal pulley (Riordan) - loop a segment of FCU around itself (Fig. 125).
　　　　　　　(Watson) - weave a segment of ECU to FCU (Fig. 126): leaves clean bed in loop, prevents radial migration of loop.

KIDS

Better to use ADQ or EDQ (need to dissect up to ulna) (Fig. 127). This does not need pulley, and therefore, do not need to worry about growth center of ulna.

NOTES

NOTES

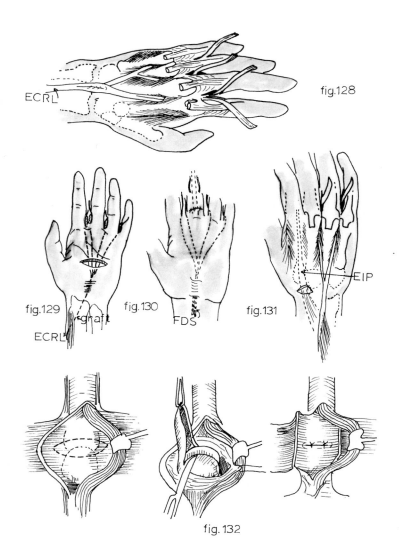

ULNAR

INTRINSIC TRANSFER

Brand[50,51]   - ECRL (with PL graft) ⟶ Rad. Lat. Band
          I - dorsally, split interossei, and under TMCL (ulnar side of index finger because no TMCL on radial side) (Fig. 128).

         II - from volar forearm, through carpal tunnel, through lumbrical canals to radial side of each finger (Fig. 129).

Bunnel[52]   - FDS (4 slips) ⟶ Rad. Lat. Band or dorsal ext. aponeurosis

(pass as in Brand II) (Fig. 130)

Fowler[53]   - EIP (2 slips) ⟶ index, middle
                 EDQ (2 slips) ⟶ ring, little
                 (all under TMCL) (Fig. 131)

Riordan[54]   - EIP (2 slips) ⟶ ring, little
                  PL (plus graft) ⟶ index, middle

CAPSULODESIS

Zancolli[55]   - only if extensors are weak, otherwise the extensors will stretch out the volar plates (Fig. 132).

INDEX ABDUCTION

EIP ⟶ insertion of 1st DI.

THUMB ADDUCTION

FCU (PL graft) ⟶ ADD.P. (Fig. 133)
Pass around hamate under hypothenar muscles in palmar space across palm under arterial arch to tendinous insertion of ADD.P.

NOTES

NOTES

236 THE HAND BOOK

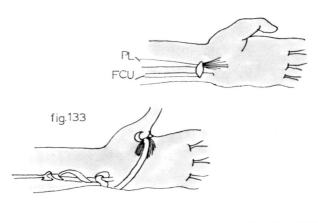

fig. 133

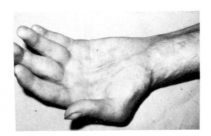
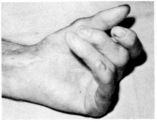

fig. 134

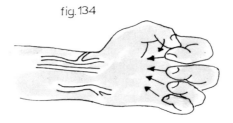

Volkmann's Contracture

VOLKMANN'S CONTRACTURE

Most frequent cause is a supracondylar fracture of humerus in child--causes vascular obstruction within the closed proximal osseofascial compartment, ISCHEMIC MUSCLE NECROSIS, and ULNAR NERVE and MEDIAN NERVE STRANGULATION compression. May be due to swelling and compression in a cast, hemorrhage in muscle bellies, secondary to infusion of chemotherapy, compression of arm in a state of necrosis.

CARDINAL SIGN IS PAIN ON PASSIVE EXTENSION OF FINGERS.

May result in (Fig. 134):

1. contracted wrist and finger flexors (FIBROSIS)
2. intrinsic muscle paralysis ⎫
3. loss of sensation ⎬ median & ulnar nerve damage
4. wrist and finger extension restrained by fibrosed flexor mass

TREATMENT:

EARLY - When signs of ischemia appear (capillary refill, paraesthesias, PAIN ON PASSIVE EXTENSION OF FINGERS). All circumferential dressings must be removed, and elbow flexion relieved.

NOTES

NOTES

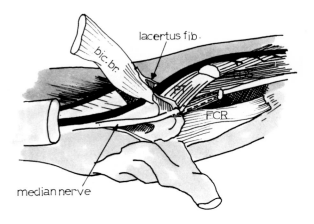

fig. 135

LATE - When fascial compartment is relieved, DISPLACE MEDIAN NERVE from within the flexor muscle group and into the subcutaneous tissue. Recovery can be expected despite compression for as long as 1 YEAR (maybe longer in children : 3-5 years).

1. resect fibrotic muscle infarct in forearm - save as much as you can, some fibers may be functional.

2. free median nerve as it passes between the two heads of the pronator teres (Fig. 135).

3. free ulnar nerve as it passes between the two heads of the FCU.

4. support wrist in extension, hand in SAFE POSITION.

CONTRACTURE STATE

1. Synnergistic tendon transfers

    ECRL ─────▶ tethered FDP

    BR ──────▶ tethered FDP

    ADQ ─────▶ APB & OP

2. May perform FLEXOR SLIDE. At 3-6 months after release, degree of return can be evaluated. For flexor slide, open upper forearm to release and slide flexors 3 inches. This may result in loss of elbow flexion, the cause of which is not very clear.

NOTES

NOTES

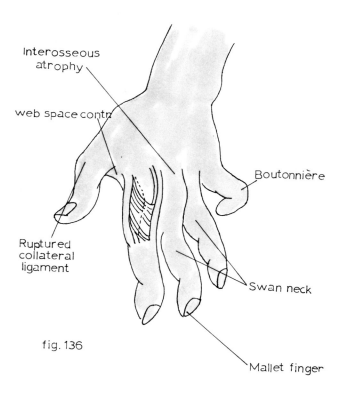

fig. 136

X
Rheumatoid Arthritis

There are two components to this disease:

1. rheumatoid synovium
2. deforming force

MP — deforming force is from contraction of intrinsic muscles resulting in volar pull and ulnar pull.

Interossei hold the dorsal hood, and ulnar and volar pull cause radial side to give.

IP — deforming force causes all types of deformities (Fig. 136):

1. Swan neck (disease of volar plate)
2. Mallet finger (disease of extensor insertion)
3. Boutonniere (disease of central slip)
4. Ruptured collateral ligament
5. Subluxation
6. Fusion

Thumb — The deforming forces are still intrinsics which cause subluxation of MP. This results in hyperextension of IP by intrinsics. When the MP subluxation is repaired and stabilized, the recurvatum at the IP corrects itself.

Treatment of MP deformities:

1. If volar subluxation of MP can be reduced, perform PRESERVATION arthroplasty. Centralize extensor tendon, plicate radial side of hood, release ulnar intrinsics, transfer tendons to ulnar band.

2. If subluxation cannot be reduced, perform RECON-STRUCTIVE arthroplasty (either Tupper or prosthetic).

NOTES

NOTES

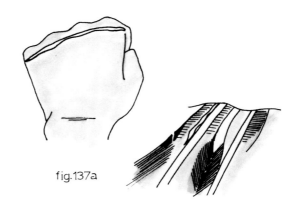

fig. 137a

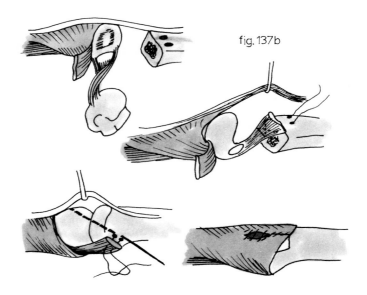

fig. 137b

RECONSTRUCTIVE ARTHROPLASTY (Fig. 137A)

1. Transverse dorsal metacarpal incision.
2. Release radial extensor hood.
3. Synovectomy of MP joint.
4. Cut intrinsic muscle insertions and resect portions proximally.
5. Release volar plate up along flexors (be careful of NV bundle).
6. Cut metacarpal heads (sagittal saw) with slight volar and radial tilt.

TUPPER[56] (Fig. 137B):

7. Drill 2 holes in metacarpal neck, wire-suture volar plate with single Bunnell X, pass into metacarpal shaft, and through dorsal holes.
8. K-wire from proximal phalanx through metacarpal shaft, with joint in extension and slight radial deviation. Then tie volar plate sutures down.
9. Reef up extensor hood on radial side.
10. Release EIP and EDQ at MP area, split tendons up into wrist, and suture to radial lateral bands.
11. Fuse thumb MP joint if necessary.
12. SPLINT X 1 WEEK - EARLY MOTION.

NOTES

NOTES

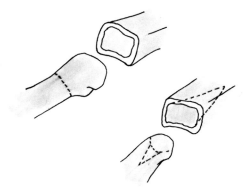

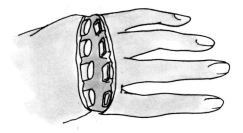

fig. 138

PROSTHETIC (Fig. 138):

1. Rongeur articular surfaces of joint and curette cavities to accept sizers.
2. Place prosthesis in each joint.
3. Reef up extensor hood on radial side.
4. Release EIP and EDQ at MP area, split tendons up into wrist and suture to radial lateral bands.
5. Fuse thumb MP joint if necessary.
6. SPLINT X 1 WEEK - EARLY MOTION.

NOTES

NOTES

NOTES

XI
Degenerative Arthritis

Degenerative arthritis may be a result of trauma, or it may be due to non-specific changes in the joint over a period of time.

Patients with CMC arthritis complain of chronic discomfort or pain on stress of this joint (e.g. opening a car door by pressing button with thumb).

Indication for arthrodesis is a need for stability for power, as in laborers. Indication for prosthetic implant is more need for motion and less for power, as in older women. The problem with implant arthroplasty is that it may sublux volarward.

NOTES

NOTES

260 THE HAND BOOK

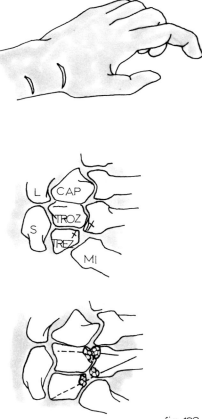

fig. 139

LIMITED ARTHRODESIS OF WRIST

Watson Technique

NEVER FUSE WRIST COMPLETELY, except for significant pain and deformity. For painful degenerative arthritis, a limited wrist fusion (of only the bones that appear to be involved by x-ray) serves the purpose of relieving symptoms, yet allows significant motion of the wrist (Fig. 139).

1. Rongeur out the cartilagenous surfaces of the joints to be fused.
2. Obtain cortical and cancellous bone grafts from radius of same hand through a second incision.
3. Pack cancellous bone into surfaces and K-WIRE.
4. Apply cortical bone into wedge and K-WIRE again.
5. SPLINT X 9 WEEKS.

   Post-op schedule:

2 weeks	-	Change dressings - sutures out - apply long arm cast.
4-6 weeks	-	Remove long arm cast and apply short arm cast.
5-7 weeks	-	Remove pins.
9 weeks	-	Remove splint and begin motion.

NOTES

NOTES

264 THE HAND BOOK

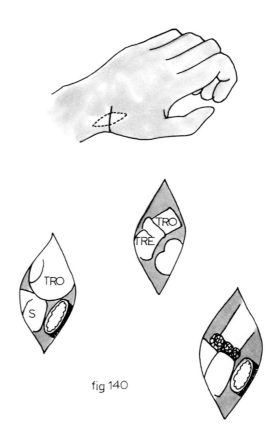

fig 140

CMC IMPLANT ARTHROPLASTY (Fig. 140)

1. Transverse incision over CMC.
2. Expose, retract EPB, make longitudinal incision in deeper structures.
3. Expose bed of first extensor compartment to get to CMC joint.
4. Strip ligaments of CMC WITH PERIOSTEUM or part of ligaments.
5. Rongeur MC base.
6. Rongeur TRAPEZIUM out, leaving ligaments on undersurface of trapezium.
7. Excise part of trapezoid over scaphoid to get good purchase of scaphoid.
8. Drill into MC shaft for sizer, cut ADD.P. off the thumb.
9. Place prosthesis in and pass 2 K-wires from shaft of MC to carpals to hold thumb in ABDUCTION.
10. GAUNTLET CAST FOR 5-6 WEEKS. May remove pins at 4-6 weeks.

NOTES

NOTES

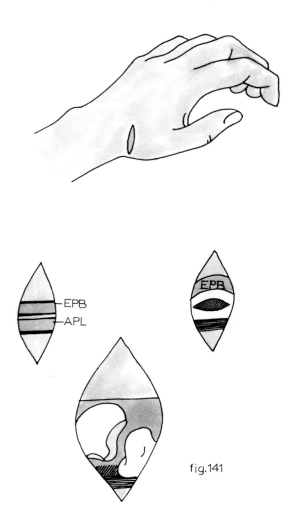

fig. 141

CMC ARTHRODESIS (Fig. 141)

1. Following exposure of CMC joint, Rongeur cartilage on surface into concave and convex surfaces.
2. RETROGRADE-PIN joint in COMPRESSION.
3. Fuse in 20° radial abduction
   35° - 40° palmar abduction
4. Close floor of sheath with absorbable suture.
5. GAUNTLET CAST FOR 5-6 weeks. May remove pins at 4-6 weeks.

NOTES

NOTES

NOTES

XII

Congenital Arthritis

Polydactyly — duplication of a digit or a part of a digit (may be skin tag only). Probably most common congenital hand anomaly--exact statistics not available.

Syndactyly — partial webbing, to complete attachment of one or more adjacent digits. Most common congenital hand anomaly if exclude minor polydactylia. Most frequent type is webbing between the second and third toes. Most common in hand is between middle and ring finger.

Brachydactyly — short fingers due to short phalanges or metacarpals.

Ectrodactyly — short fingers due to absence of one or more phalanges or metacarpals.

Symphalangism — end-to-end fusion of phalanges resulting in shortened or fused fingers. May have only MP joint function.

Camptodactyly — congenital flexion deformity of finger.

Clinodactyly — lateral deviation or angulation of finger.

Arthrogryposis — congenital flexion or angulation deformities associated with congenital muscle deficiency and joint deformities.

NOTES

NOTES

280 THE HAND BOOK

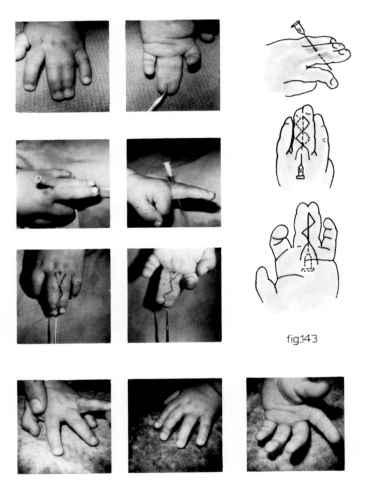

fig.143

fig 144

SYNDACTYLY

A normal web has 2 wide areas--at the dorsal MP area, and at the volar margin. Therefore, two interdigitated triangular flaps give wide blunt ends, with no suture lines at the margins to contract. Surgery may be delayed till patient is 4 years old, unless a short finger is causing angulation and deformity of next finger.

1. Pass needle from dorsal skin to level of adjacent normal dorsal web, and allow it to surface from volar skin at level of adjacent normal volar web (Fig. 143).

2. These two points are the centers of the bases of the two triangular flaps to be constructed.

3. The length of each triangular flap is the length of the needle that is buried between the dorsal and volar skin.

4. Draw zig-zag flaps on volar and dorsal skin.

5. Suture with absorbable material.

6. Apply full thickness skin graft in defects.

7. Wet cotton dressings for conformity to fingers. Plaster with elbow at 90° -- long arm cast and apply Velpeaux shoulder sling (see pg. 88).

8. Remove dressings in 8 days (Fig. 144).

NOTES

NOTES

284 THE HAND BOOK

fig.145

DUPLICATED THUMB

Both phalanges usually have FPL, but the good thumb usually has the EPL.

The FPL forks up and divides. When one phalanx is excised, centralize the FPL and stabilize collateral ligament. Otherwise there is unstable distal phalanx.

1. Reconstruct flexor tendon to CENTRALIZE.

2. Flip RPL over from radial phalanx to ulnar side of preserved phalanx (Fig. 145).

3. EIP or free graft to form EPL.

4. The IP stability may correct once tendon is centralized.

5. SPLINT X 3 WEEKS.

NOTES

NOTES

NOTES

NOTES

NOTES

NOTES

Artwork Acknowledgements

Several of the illustrations in this book were modified from previously published works by permission. Specific credit is hereby given accordingly:

Burton RI: Severed tendons and nerves distal to metacarpal joint. In Littler JW, Cramer LM, Smith JW, editors: Symposium on Reconstructive Hand Surgery, Vol IX, St. Louis, 1974, The C.V. Mosby Co., (for Fig. 15-17, 20, 43).

Burton RI, Littler JW: Nontraumatic Soft Tissue Afflictions of the Hand. In Ravitch MM, et al. (eds): Current Problems in Surgery. Chicago, 1975, Year Book Medical Publishers (for Fig. 87-89, 91).

Chase RA: Atlas of Hand Surgery. Philadelphia, 1973, W.B. Saunders Co. (for Fig. 43-46, 48, 52-56, 58, 85, 101, 111, 112, 122, 123, 125, 127-132).

Eaton RG: Joint Injuries of the Hand. Springfield, 1971, Charles C. Thomas (for Fig. 6, 8, 11, 12, 14).

Flynn JE: Hand Surgery. (2nd edition) Baltimore, 1975, Williams and Wilkins (for Fig. 95, 96).

Grabb WC, Smith JW: Plastic Surgery. A Concise Guide to Clinical Practice. 2nd edition. Boston 1973, Little Brown and Company (for Fig. 60).

Gray H: Anatomy of the Human Body, Goss CM, editor, (29th edition) Philadelphia 1973, Lea & Febiger (for Fig. 32-41).

Johnson FG, Green MH: Another cause of irreducible dislocation of the proximal interphalangeal joint of a finger. J Bone Joint Surg 48A:542, 1966 (for Fig. 74).

Kahn S: A dynamic tenodesis of the distal interphalangeal joint for use after severances of the profundus alone. Plast Reconstr Surg 51:536, 1973. Williams and Wilkins (for Fig. 108).

Kaplan EB: Dorsal dislocation of the metacarpophalangeal joint of the index finger. J Bone Joint Surg 39A: 1081, 1957 (for Fig. 75).

Littler JW: Principles of reconstructive surgery of the hand. In Converse JM, editor. Reconstructive Plastic Surgery (1st edition) Philadelphia 1964, W.B. Saunders Company (for Fig. 21-23).

Littler JW: Principles of reconstructive surgery of the hand. The digital extensor-flexor system. In Converse JM, editor, Reconstructive Plastic Surgery (2nd edition) Philadelphia 1977, W.B. Saunders Co. (for Fig. 3, 4, 20, 135).

Littler JW: Hand structure and function. In Littler JW, Cramer LM, Smith JW, editors: Symposium on Reconstructive Hand Surgery, Vol IX, St. Louis, 1974, C.V. Mosby Co. (for Fig. 1,2).

McFarlane RM: Patterns of the diseased fascia in the fingers in Dupuytren's contracture. Plast Reconstr Surg 54:31, 1974. Williams and Wilkins (for Fig. 92).

Milford L: The Hand St. Louis 1971, C.V. Mosby Company (for Fig. 71, 72).

Bibliography

1. Littler JW: Hand structure and function. Symposium on Reconstructive Hand Surgery, St. Louis, CV Mosby Co., 1974.

2. Eaton RG: Joint Injuries of the Hand. Springfield, Charles C. Thomas, 1971.

3. Landsmeer JMF: The anatomy of the dorsal aponeurosis of the human finger and its functional significance. Anat Rec 104:31, 1941.

    Landsmeer JMF: The coordination of finger-joint motions. J Bone Joint Surg 45A:1654, 1963.

4. Shrewsburg MM, Johnson RK: A systematic study of the oblique retinacular ligament of the human finger: Its structure and function. J Hand Surg 2:194, 1977.

5. Furlow LT: Cause and prevention of tourniquet ooze. Surg Obstet Gynecol 132:1069, 1971.

6. Laing PG: Vascular anatomy and shunting systems through the humerus. Symposium on Reconstructive Hand Surgery. Saint Louis, CV Mosby Co., 1974, p. 55.

7. Froment J: Presse Med. Oct. 21, 1915.

8. Phalen GS, Kendrick JI: Compression neuropathy of the median nerve in the carpal tunnel. JAMA 164:524, 1957.

9. Hamer ML, Robson MC, Krizek TJ, et al: Quantitative bacterial analysis of comparative wound irrigations. Ann Surg 181:819, 1975.

10. Rodeheaver GT, Pettry D, Thacker JG: Wound cleansing by high pressure irrigation. Surg Gynecol Obstet 141:357, 1975.

11. Bruner JM: The zig-zag volar digital incision for flexor tendon surgery. Plast Reconstr Surg 40:571, 1967.

12. Deming EG: Y-V advancement pedicles in surgery for Dupuytren's contracture. Plast Reconstr Surg 29: 581, 1962.
13. Browne EZ, Dunn HK, Snyder CC: Ski pole thumb injury. Plast Reconstr Surg 58:19, 1976.
14. Parikh M, Nahigian S, Froimson A: Gamekeeper's thumb. Plast Reconstr Surg 58:24, 1976.
15. Kutler W: A new method for finger tip amputation. JAMA 133:29, 1947.
16. Atasoy E, Iokimidis E, Kasden ML, Kutz JE, Kleinert HE: Reconstruction of the amputated finger tip with a triangular volar flap. J Bone Joint Surg 52A:921, 1970.
17. Moberg E: Aspects of sensation in reconstructive surgery of the upper extremity. J Bone Joint Surg 46A:817, 1964.
18. Snow JW: The use of a volar flap for repair of finger tip amputations: a preliminary report. Plast Reconstr Surg 40:163, 1967.
19. Flagg SV, Finseth FJ, Krizek TJ: Ring avulsion injury. Plast Reconstr Surg 59:241, 1977.
20. Isebin F, Levame J, Godoy J: A simplified technique for treating mallet fingers: Tenodermodesis. J Hand Surg 2:118, 1977.
21. Ariyan S, Watson HK: The palmar approach for the direct visualization and release of the carpal tunnel. Plast Reconstr Surg 60:539, 1977.
22. Graham WP: Variations of the motor branch of the median nerve at the wrist. Plast Reconstr Surg 51:90, 1973.
23. Lanz U: Anatomical variations of the median nerve in the carpal tunnel. J Hand Surg 2:44, 1977.
24. McFarlane RM: Patterns of the diseased fascia in the fingers in Dupuytren's contracture. Plast Reconstr Surg 54:31, 1974.
25. McCash CR: The open palm technique in Dupuytren's contracture. Brit J Plast Surg 17:271, 1964.

BIBLIOGRAPHY

26. Ariyan S, Krizek TJ: In defense of the open wound. Arch Surg 111:293, 1976.

27. Ariyan S: A simple stereotactic method to isolate and remove foreign bodies. Arch Surg 112:857, 1977.

28. Kanavel AB: Study of acute phlegmons of the hand. Surg Gynecol Obstet 1:221, 1905.

29. Kanavel AB: INFECTIONS OF THE HAND. A GUIDE TO THE SURGICAL TREATMENT OF ACUTE AND CHRONIC SUPPURATIVE PROCESSES IN THE FINGERS, HAND, AND FOREARM. 7th edition. Philadelphia, Lea & Febiger, 1939.

30. Carter SJ, Mersheimer W: Infections of the hand. Orthop Clin North Am 1:455, 1970.

31. Arons MS: Purposeful delay of the primary repair of cut flexor tendons in "Some-Man's-Land", in children. Plast Reconstr Surg 53:638, 1974.

32. Hunter JM, Salisbury RE: Use of gliding implants to produce tendon sheaths. Plast Reconstr Surg 45:564, 1970.

33. Farkas LG, McCain WG, Sweeney P, Wilson W, Hurst LN, Lindsay WK: An experimental study of the changes following silastic rod preparation of a new tendon sheath and subsequent tendon grafting. J Bone Joint Surg 55A:1149, 1973.

34. Kahn S: A dynamic tenodesis of the distal interphalangeal joint for use after severances of the profundus alone. Plast Reconstr Surg 51:536, 1973.

35. Leddy JP, Packer JW: Avulsion of the profundus tendon insertion in athletes. J Hand Surg 2:66, 1977.

36. Reynolds B, Wray RC, Weeks PM: Should an incompletely severed tendon be sutured? Plast Reconstr Surg 57:36, 1976.

37. Kleinert HE: Response to Reynolds, Wray, Weeks: Plast Reconstr Surg 57:36, 1976.

38. Wray RC, Holtmann B, Weeks PM: Clinical treatment of partial tendon lacerations without suturing and with early motion. Plast Reconstr Surg 59:231, 1977.

39. Watson HK, Ritland GD: Post-traumatic interosseus-lumbrical adhesions. J Bone Joint Surg 56A: 79, 1974.

40. Wheeldon FT: Recurrent dislocation of extensor tendons in the hand. J Bone Joint Surg 36B:612, 1954.

41. Elson RA: Dislocation of the extensor tendons of the hand. J Bone Joint Surg 49B:324, 1967.

42. McCoy FJ, Winsky AJ: Lumbrical loop operation for luxation of the extensor tendons of the hand. Plast Reconstr Surg 44:142, 1969.

43. Millesi H, Meissl G, Berger A: The interfascicular nerve grafting of the median and ulnar nerves. J Bone Joint Surg 54A:727, 1972.

44. Millesi H: Microsurgery of peripheral nerves. Hand 5: 157, 1973.

45. Cabaud HE, Rodkey WG, McCarroll HR, Mutz SB: Epineurial and perineurial fascicular nerve repairs: a critical comparison. J Hand Surg 1:131, 1976.

46. Bora FW, Pleasure DE, Didizian NA: A study of nerve regeneration and neuroma formation after nerve suture by various techniques. J Hand Surg 1:138, 1976.

47. Dellon AL, Curtis RM, Edgerton MT: Reeducation of sensation in the hand after nerve injury and repair. Plast Reconstr Surg 53:297, 1974.

48. Royle ND: An operation for paralysis of the intrinsic muscles of the thumb. JAMA 111:612, 1938.

49. Thompson TC: Modified operation for opponens paralysis. J Bone Joint Surg 24:632, 1942.

50. Brand PW: Paralytic claw hand. With special reference to paralysis in leprosy and treatment by the sublimis transfer of Stiles and Bunnell. J Bone Joint Surg 40-B:618, 1958.

51. Brand PW: Tendon grafting. Illustrated by a new operation for intrinsic paralysis of the fingers. J Bone Joint Surg 43B:444, 1961.

BIBLIOGRAPHY

52. Bunnell S: Tendon transfers in the hand and forearm. American Academy of Orthopedic Surgeons Instructional Course Lectures, Vol. 6. Ann Arbor, J.W. Edwards, 1949.

53. Fowler B quoted by Riordan DC: Tendon transplantation in median and ulnar-nerve paralysis. J Bone Joint Surg 53A:312, 1953.

54. Riordan DC: Surgery of the paralytic hand. American Academy of Orthopedic Surgeons Instructional Course Lectures Vol. 16, St. Louis, C.V. Mosby Co., 1959.

55. Zancolli EA: Claw hand caused by paralysis of the intrinsic muscles. A simple surgical procedure for its correction. J Bone Joint Surg 39A: 1076, 1957.

56. Tupper JW: quoted by Flatt AE: <u>The Care of the Rheumatoid Hand.</u> St. Louis, C.V. Mosby Co., 1974, p. 188.

Index

adductor transfer	233
Allen test	65
anatomy	
arterial	53
bone	1
joints	5
ligaments	9
muscles	33
nerves	41, 209
tendons	17
arthritis	
degenerative	257
rheumatoid	245
arthrodesis (fusion)	261
CMC	269
scaphoid	273
arthrogryposis	277
arthroplasty	
implant	265
prosthetic	253
reconstructive	249
Tupper	249
assembly line	9
Bennett fracture	97
brachdactyly	277
Brand transfers	233
Boutonniere	
deformity	125
test	65
Bunnell	
stitch	177
test	65

INDEX

camptodactyly	277
capsulectomy	161
capsulodesis	233
carpal tunnel	13
syndrome	141
check-rein ligaments	9
clinodactyly	277
collateral ligaments	9
congenital deformities	277
contractures	
extension	161
flexion	153
DeQuervain's disease	137
dislocations	101,105,205
duplication thumb	285
Dupuytren's disease	149
ectrodactyly	277
examination	61
fingertip injury	113
Finkelstein test	65
flexor	
tendon injury	173
"slide"	241
foreign bodies	165
Fowler	
test	65
transfer	233
fractures	93,97
free tendon graft	185
Froment's sign	69
Gamekeeper's thumb	109
graft, tendon	185
Guyon's canal	13
Hunter rod	177

300

INDEX

incisions	77
infection	169
intrinsic	
muscles	37
transfers	233
Landsmeer ligaments	25
ligaments	
anatomy	9
of Landsmeer	25
tear	101
limited arthrodesis	261
Littler	
incision	77
release	161
tenodesis	129
mallet finger	121
muscles	
extrinsic	33
intrinsic	37
nerve	
anatomy	41,209
blocks	69
digital	213
repair	209
supply	49
neurovascular island	217
non-specific tenosynovitis	133
palmaris longus	189
Phalen's test	69
polydactyly	277
pulleys, tendon	189
recurvatum deformity	129

INDEX

saddle deformity	197
sagittal bands	21
silastic sheet	205
snuffbox	29
splinting	
technique	85
duration	89
stenosing tenosynovitis	133
steroid injection	133
suppurative tenosynovitis	169
swan neck deformity	129
symphalangism	277
syndactyly	277, 281
tendon	
advancement	181
dislocation	205
extensor injury	201
flexor injury	173
graft	185
partial laceration	193
pulleys	189
palmaris longus	189
rupture	193
transfers	221
weave	177
tenodermodesis	121
tenodesis	
flexor	129
Littler	129
profundus	189
sublimis	193
tenosynovitis	
non-specific	133
stenosing	133
suppurative	169
tests	
Allen	65
Boutonniere	65
Bunnell	65
Finkelstein	65
Fowler	65

transfers	233
adductor	233
Brand	233
Bunnell	233
Fowler	233
intrinsic	233
Riordan	233
tendon	221
Zancolli	233
transverse carpal ligament	13
triangular ligament	25
trigger finger	133
Tupper arthroplasty	249
volar carpal ligament	13
volar plate	9
Volkmann's contracture	237
Watson	
limited arthrodesis	261
flexor pulley	189
opposition pulley	229
Y-V plasty	81
Z-plasty	77